T. A. Springer

Linear Algebraic Groups

1981

Birkhäuser
Boston · Basel · Stuttgart

Author

T.A. Springer
Mathematical Institute
Budapestlaan 6
Utrecht 3584 CD
Netherlands

Library of Congress Cataloging in Publication Data
Springer, Tonny Albert, 1926-
Linear algebraic groups.
(Progress in mathematics; 9)
Bibliography: p.
Includes index.
1. Linear algebraic groups. I. Title. II. Series: Progress in mathematics (Cambridge, Mass.); 9.
QA171.S718 512′.55 80-29221
3-7643-3029-5

CIP — Kurztitelaufnahme der Deutschen Bibliothek
Springer, Tonny A.:
Linear algebraic groups/T.A. Springer.
Boston; Basel; Stuttgart: Birkhäuser, 1980.
(Progress in mathematics; 9)
ISBN 3-7643-3029-5 NE: GT

3-7643-3029-5
Printed in USA

Contents.

Introduction.

These notes contain an introduction to the theory of linear algebraic groups over an algebraically closed ground field. They lead in a straightforward manner to the basic results about reductive groups.

The main difference with the existing introductory texts on this subject (e.g. those of Borel and Humphreys) lies in the treatment of the prerequisites from algebraic geometry and commutative algebra. These texts assume a number of such prerequisites, whereas I have tried to give proofs of everything. I have also tried to limit as much as possible the commutative algebra. For example, the use of the concept of normality has been avoided.

Moreover, in the later chapters most of the facts about root systems which are needed are proved, in an ad hoc manner. Exception must be made for the results on classification of root systems which are used in the last two chapters.

The exercises contain additional material. Sometimes use is made later on of the results contained in the easier exercises.

Except for a brief discussion in chapter 3 of groups over finite fields, these notes do not contain material about algebraic groups over non algebraically closed ground fields. An adequate treatment of such material would probably require another volume of the same size.

The notes had their origin in a course on linear algebraic groups, given at the University of Notre Dame in the fall of 1978. This course covered most of the material contained in the first ten chapters. I have added two chapters, with a treatment of the uniqueness and existence theorems for reductive groups.

I am grateful to my colleagues at the University of Notre Dame, in particular O.T. O'Meara and W.J. Wong, for their invitation to give a course on algebraic groups. I am also grateful to Hassan Azad, for making a first draft of the notes and to F.D. Veldkamp, for a thorough critical reading of the manuscript. Finally I want to thank Renske Kuipers for the efficient preparation of the manuscript.

Utrecht, October 1980. T.A. Springer.

1. Some algebraic geometry.

1.1. The Zariski topology.

1.1.1. Let k be an algebraically closed field, and put $V = k^n$. The elements of the polynomial algebra $S = k[T_1,\ldots,T_n]$ (abbreviated to k[T]) can be viewed as k-valued functions on V. We say that $v \in V$ is a zero of $f \in k[T]$ if $f(v) = 0$. Moreover, v is a zero of the ideal I of S if $f(v) = 0$ for all $f \in I$. We denote by $\mathcal{V}(I)$ the set of zeros of the ideal I. If X is any subset of V, let $\mathcal{I}(X) \subset S$ be the ideal formed by the $f \in S$ vanishing on X.

Recall that the radical $\sqrt{I}$ of the ideal I is the ideal of all $f \in S$ such that $f^n \in I$ for some integer $n \geq 1$. A radical ideal is one equal to its radical. It is obvious that all $\mathcal{I}(X)$ are radical ideals.

We shall need Hilbert's Nullstellensatz, in two (equivalent) formulations.

1.1.2. Theorem ("Nullstellensatz") (i) *If I is a proper ideal of S then* $\mathcal{V}(I) \neq \phi$;
(ii) *For any ideal I of S we have* $\mathcal{I}(\mathcal{V}(I)) = \sqrt{I}$.
For a proof see, for example [26, Ch. X, §2]. We also give a proof in the appendix to this chapter.

1.1.3. *Zariski topology on V.*
The function $I \mapsto \mathcal{V}(I)$ on ideals of S has the following properties:
(a) $\mathcal{V}(\{0\}) = V$, $V(S) = \phi$;
(b) *If* $I \subset J$ *then* $\mathcal{V}(J) \subset \mathcal{V}(I)$;

(c) $\mathcal{V}(I \cap J) = \mathcal{V}(I) \cup \mathcal{V}(J)$;

(d) *If* $(I_\alpha)_{\alpha\in A}$ *is a family of ideals and* $I = \sum_{\alpha\in A} I_\alpha$ *their sum then* $\mathcal{V}(I) = \bigcap_{\alpha\in A} \mathcal{V}(I_\alpha)$.

The proof of these properties is left to the reader (hint for (c): use that $IJ \subset I \cap J$).

It follows from (a), (c) and (d) that there is a topology on V whose closed sets are the $\mathcal{V}(I)$, I running through the ideals of S. This is the *Zariski topology*. The induced topology on a subset X of V is the Zariski-topology of X. A closed set in V is called an *algebraic set*.

1.1.4. *Exercises*. (1) Let V = k. The proper algebraic sets are the finite ones.

(2) Let X be any subset of V. Its Zariski closure is $\mathcal{V}(\mathcal{I}(X))$.

(3) The map $\mathcal{I}$ defines an order-reversing bijection of the family of Zariski-closed subset of V onto the family of radical ideals of S, its inverse is $\mathcal{V}$.

(4) The Euclidean topology on $\mathbb{C}^n$ is finer than the Zariski topology.

1.1.5. *Proposition. Let* $X \subset V$ *be an algebraic set.*

(i) *The Zariski topology of X is* T_1, *i.e. points are closed;*

(ii) *Any family of closed subsets of X contains a minimal one;*

(iii) *If* $X_1 \supset X_2 \supset \ldots$ *is a descending sequence of closed subsets of X there is h such that* $X_i = X_h$ *for* $i > h$;

(iv) *Any open covering of X has a finite subcovering.*

If $x = (x_1,\ldots,x_n) \in X$ then x is the zero of the ideal of S generated by $T_1-x_1,\ldots,T_n-x_n$. This implies (i).

(ii) and (iii) follow from the fact that S is a noetherian ring

[26, Ch. VI, §1], using 1.1.4(3).

To establish (iii) we formulate it in terms of closed sets. We then have to show: if $(I_\alpha)_{\alpha \in A}$ is a family of ideals such that $\bigcap_{\alpha \in A} V(I_\alpha) = \phi$, then already a finite intersection of some $V(I_\alpha)$ is empty. Now using properties (a), (d) of 1.1.3 and 1.1.4(3) we have $\sum_{\alpha \in A} I_\alpha = S$. Hence there are finitely many of the I_α, say $I_1, I_2, \ldots, I_h$ such that $I_1 + \ldots + I_h = S$. Then $\bigcap_{i=1}^{h} V(I_i) = \phi$.

A topological space X with the property (ii) is called *noetherian*. Notice that (ii) and (iii) are equivalent properties (compare the corresponding properties in noetherian rings, cf. [26, p. 142]). X is *quasi-compact* if it has the property of (iv).

1.1.6. *Exercise*. A closed subset of a noetherian space (with the induced topology) is noetherian.

1.2. Irreducibility of topological spaces.

1.2.1. A topological space X is *reducible* if it is the union of two proper closed subsets. Otherwise X is *irreducible*. A subset $A \subset X$ is irreducible if it is irreducible for the induced topology.
Notice the following fact: X is irreducible if and only if any two non-empty open subsets of X have a non-empty intersection.

1.2.2. *Exercise*. An irreducible Hausdorff space is reduced to a point.

1.2.3. *Lemma*. *Let X be a topological space.*

(i) $A \subset X$ *is irreducible if and only if its closure* $\bar{A}$ *is irreducible*;

(ii) *Let* $f: X \to Y$ *be a continuous map. If* X *is irreducible then so is* fX.

Let A be irreducible. If $\bar{A}$ is the union of two closed subsets A_1 and A_2, then A is the union of the closed subsets $A \cap A_1$ and $A \cap A_2$, whence (say) $A \cap A_1 = A$, and $A \subset A_1$, $\bar{A} \subset A_1$. Hence $\bar{A} = A_1$. So $\bar{A}$ is irreducible.

Conversely, if $\bar{A}$ is irreducible, and if A is the union of two closed subsets $A \cap B_1$, $A \cap B_2$, where B_1, B_2 are closed in X, then $\bar{A} \subset B_1 \cup B_2$. So $\bar{A} \cap B_1 = \bar{A}$ (say), whence $A \cap B_1 = A$. The irreducibility of A follows.

The proof of (ii) is easy and can be omitted.

1.2.4. *Proposition. Let* X *be a noetherian topological space.*

(i) X *is a union of finitely many irreducible closed subsets, say* $X = X_1 \cup \ldots \cup X_s$;

(ii) *If there are no inclusions among the* X_i, *they are uniquely determined, up to order.*

Recall that a noetherian space is one with the property of 1.1.5(ii). If (i) is false, the noetherian property and 1.1.6 show that there is a minimal closed subset A of X which is *not* a finite union of irreducible closed subsets. Then A must be reducible, so A is a union of two proper closed subsets. But these do have the property in question, and a contradiction emerges. This establishes (i).

To prove (ii), assume there are no inclusions among the X_i, and let $X = Y_1 \cup \ldots \cup Y_t$ be a second decomposition with the same properties. Then $X_i = \bigcup_i (X_i \cap Y_j)$, and by the irreducibility there is a function $f: \{1,\dots,s\} \to \{1,\dots,t\}$ with $X_i \subset Y_{f(i)}$. Similarly, there is $g: \{1,\dots,t\} \to \{1,\dots,s\}$ with $Y_j \subset Y_{g(j)}$. Since $X_i \subset X_{g(f(i))}$, we have $g \circ f = \mathrm{id}$, also $f \circ g = \mathrm{id}$. This implies (ii).

The X_i are called the (irreducible) <u>components</u> of X. We now return to the Zariski topology on $V = k^n$.

1.2.5. <u>Proposition</u>. <u>A closed subset X of V is irreducible if and only if $I(X)$ is a prime ideal.</u>

Let X be irreducible and let $f,g \in S$ be such that $fg \in I(X)$. Then $X = (X \cap V(fS)) \cup (X \cap V(gS))$ and the irreducibility of X implies that $X \subset V(fS)$, say, i.e. $f \in J(X)$. So $J(X)$ is prime. Conversely, let $I(X)$ be prime, and let $X = V(I_1) \cup V(I_2) = V(I_1 \cap I_2)$. If $X \neq V(I_1)$, there is $f \in I_1$ with $f \notin I(X)$. Since $fg \in I(X)$ for all $g \in I_2$, it follows from the fact that $I(X)$ is prime that $I_2 \subset I(X)$, whence $X = V(I_2)$. So X is irreducible.

1.2.6. <u>Exercise</u>. (1) Let X be a noetherian space. The components of X are the maximal irreducible closed subsets of X.
(2) Any radical ideal I of S is an intersection $I = P_1 \cap \dots \cap P_s$ of finitely many prime ideals. If there are no inclusions among the P_i they are uniquely determined, up to order.

1.3. <u>Affine</u> k-<u>algebras</u>.

1.3.1. We now turn to more intrinsic descriptions of algebraic sets. Let $X \subset V$ be one. The restrictions to X of the polynomial functions of S form a k-algebra, denoted by k[X], which clearly is isomorphic to $S/I(X)$. The following properties of k[X] are obvious:

(a) k[X] _is a commutative_ k-_algebra of finite type_, i.e. there is a finite subset $\{x_1,\dots,x_r\}$ of k[X] such that $k[X] = k[x_1,\dots,x_r]$;

(b) k[X] _is reduced_, i.e. 0 is the only nilpotent element of k[X]. A k-algebra A with the properties of (a) and (b) is called an _affine_ k-_algebra_. If A is an affine k-algebra there is an r and an algebraic subset X of k^r such that $A \simeq k[X]$. For $A \simeq k[T_1,\dots,T_r]/I$, where I is the kernel of the homomorphism $k[T_1,\dots,T_r] \to A$ sending T to x_i, this is a radical ideal. We call k[X] the _affine algebra_ of X.

1.3.2. We next show that the set X together with its Zariski topology is determined by its affine algebra k[X]. First observe that (by 1.2.5) X is irreducible if and only if k[X] is an integral domain. If I is an ideal in k[X] let $V_X(I)$ be the set of the $x \in X$ such that $f(x) = 0$ for all $f \in I$. If Y is a subset of X, let $I_X(Y)$ be the ideal of the $f \in k[X]$ such that $f(x) = 0$ for all $x \in Y$.

A being an affine algebra, let Max (A) be the set of its maximal ideals. If X is as before, and $x \in X$, denote by M_x the ideal of all $f \in k[X]$ vanishing in x. Then M_x is a maximal ideal (since $k[X]/M_x$ is the field k).

1.3.3. _Proposition_. (i) _The map_ $x \mapsto M_x$ _defines a bijection of_ X _onto_ Max(k[X]), _moreover_ $x \in V_X(I)$ _if and only if_ $I \subset M_x$;
(ii) _The closed sets of_ X _are the_ $V_X(I)$, I _running through the_

ideals of k[X].

Since $k[X] \simeq S/I(X)$, the maximal ideals of k[X] correspond to the maximal ideals of S containing $I(X)$. Let M be a maximal ideal of S. Then 1.1.4(3) and 1.1.5(ii) imply that M is the set of all $f \in S$ vanishing in some $x \in k^n$. From this the first point of (i) follows, and the second point is obvious.

(ii) is a direct consequence of the definition of the Zariski topology of X.

1.3.4. _Exercises_. (1) For any ideal I of k[X] we have $I_X(V_X(I)) = \sqrt{I}$; for any subset Y of X we have $V_X(I_X(Y)) = \overline{Y}$.

(2) The map I_X defines an order-reversing bijection of the family of Zariski-closed subsets of X onto the family of radical ideals of k[X], its inverse is V_X.

(3) Let A be an affine k-algebra. Define a bijection of Max(A) onto the set of k-algebra homomorphisms $A \to k$.

From 1.3.3 we see that the algebra k[X] completely determines X and its Zariski topology.

1.3.5. We also have to consider locally defined functions on X. For this we need special open subsets of X, which we now define.

If $f \in k[X]$, put

$$D(f) = \{x \in X \mid f(x) \neq 0\}.$$

Clearly, this is an open subset, viz. the complement of $V_X(fk[X])$. It is also clear that

$$D(fg) = D(f) \cap D(g), \quad D(f^n) = D(f) \quad (n \geq 1).$$

We call the D(f) _principal open subsets_ of X.

1.3.6. _Lemma_. (i) _If_ $f,g \in k[X]$ _and_ $D(f) \subset D(g)$ _then_ $f^n \in gk[X]$ _for some_ $n \geq 1$;

(ii) _The_ D(f) _form a basis of the topology of_ X.

Using 1.1.4(3) we see that $D(f) \subset D(g)$ if and only if $\sqrt{fk[X]} \subset \sqrt{gk[X]}$, which implies (i).

To prove (ii) one has to show that any closed set is an intersection of sets of the form $V_X(fk[X])$, which is obvious from the definitions.

1.4. _Regular functions, ringed spaces_.

1.4.1. The notations are as before. Let $x \in X$. A k-valued function defined in a neighbourhood U of x is called _regular in_ x if there are $g,h \in k[X]$ such that $h(x) \neq 0$ and such that there is an open neighbourhood $V \subset U$ of x with $h(y) \neq 0$ and $f(y) = g(y)h(y)^{-1}$ for all $y \in V$.

A function f defined in a non-empty open subset U of X is _regular_ if it is regular in all points of U (so for each $x \in X$ there exist g_x, h_x with the properties stated above, they may depend on x). We denote by $\mathcal{O}_X(U)$ or $\mathcal{O}(U)$ the k-algebra of regular functions in U.

The following properties are obvious:

(A) _If_ V _is a non-empty open subset of the open set_ U, _the restriction map of functions defines a_ k-_algebra homomorphism_ $\mathcal{O}(U) \to \mathcal{O}(V)$;

(B) _Let_ $U = \bigcup_{\alpha \in A} U_\alpha$ _be an open covering of the open set_ U. _Suppose that for each_ $\alpha \in A$ _we are given_ $f_\alpha \in \mathcal{O}(U_\alpha)$ _such that_

f_α <u>and</u> f_β <u>restrict to the same function on</u> $U_\alpha \cap U_\beta$ $(\alpha,\beta \in A)$. <u>Then there is</u> $f \in \mathcal{O}(U)$ <u>whose restriction to</u> U_α <u>is</u> f_α, <u>for all</u> $\alpha \in A$.

1.4.2. <u>Sheaves of functions</u>.

Let X be a topological space, and suppose for each non-empty open subset U of X a k-algebra $\mathcal{O}(U)$ is given such that (A) and (B) hold. Then $\mathcal{O}$ is a <u>sheaf of</u> k-<u>valued functions</u> on X (we shall not need the general notion of a sheaf on a topological space). A pair $(X,\mathcal{O})$ of a topological space and a sheaf of functions is called a <u>ringed space</u>.

Let $(X,\mathcal{O})$ be a ringed space, let Y be a subset of X. We define the induced ringed space $(Y,\mathcal{O}|Y)$ as follows. Y is provided with the induced topology. If U is an open subset of Y, then $(\mathcal{O}|Y)(U)$ consists of the functions f on U with the following property: there exists an open covering $U \subset \bigcup_{\alpha\in A} U_\alpha$ by open sets of X and for each $\alpha \in A$ an element $f_\alpha \in \mathcal{O}(U_\alpha)$ whose restriction to $U_\alpha \cap U$ coincides with that of f.

Then $\mathcal{O}|Y$ is a sheaf of functions on Y. The proof of this fact is left to the reader. If Y is open in X, then $(\mathcal{O}|Y)(U)=\mathcal{O}(U)$ for all open $U \subset Y$.

1.4.3. <u>Affine algebraic varieties</u>.

The ringed spaces $(X,\mathcal{O}_X)$ of 1.4.1. are the <u>affine algebraic varieties</u> over k. In the sequel we shall usually drop the $\mathcal{O}_X$, and speak of an algebraic variety X,... . In this case, denote by $\mathcal{O}_{x,X}$ the k-algebra of functions regular in $x \in X$. By definition these are functions defined and regular in some open neighbourhood of x, two such functions being identified if they coincide in some neighbourhood of x (a formal definition is: $\mathcal{O}_{X,x} = \varinjlim \mathcal{O}_X(U)$, U running through the open neighbourhoods of x, ordered by inclusion and $\varinjlim$ denoting inductive limit). We write $\mathbb{A}^n$ for the affine variety k^n, this is affine n-space.

1.4.4. Exercises. (1) $\mathcal{O}_{X,x}$ is a local ring, i.e. has only one maximal ideal (viz. the ideal of "functions vanishing in x").
(2) Let $M_x \subset k[X]$ be the ideal of functions vanishing in x. Show that $\mathcal{O}_{X,x}$ is isomorphic to the localization $k[X]_{M_x}$ (If A is a commutative ring and S a multiplicatively closed subset of A, the ring of fractions $S^{-1}A$ is the quotient of $A \times S$ by the equivalence relation: $(a,s) \sim (a',s')$ if and only if there is $s'' \in S$ with $s''(s'a-sa') = 0$. The equivalence class of (a,s) is written as a fraction $\frac{a}{s}$, and these are added and multiplied in the usual way. If P is a prime ideal in A and S=A-P, then $S^{-1}A$ is written A_P, and is called the localization of A at P. See [26 , Ch. II, §3]).

Let $(X,\mathcal{O}_X)$ be an affine algebraic variety. If follows from the definitions that there is a homomorphism ϕ: $k[X] \to \mathcal{O}_X(X)$, namely the identity map on functions.

1.4.5. Theorem. *ϕ is an isomorphism.*
Injectivity is obvious, so we have to prove surjectivity. Let $f \in \mathcal{O}_X(X)$. For each $x \in X$ there exist an open neighbourhood U_x of x and $g_x, h_x \in k[X]$ such that h_x does not vanish on U_x and that

$$f(y) = g_x(y)h_x(y)^{-1} \quad (y \in U_x).$$

By 1.3.6(ii) we may assume that there is $a_x \in k[X]$ with

$U_x = D(a_x)$. Then $D(a_x) \subset D(h_x)$, and by 1.3.6(i) there exist $h'_x \in k[X]$ and $n_x \geq 1$ with

$$a_x^{n_x} = h_x h'_x.$$

Then on U_x we have $f = g_x h'_x (a_x^{n_x})^{-1}$. Observing that $D(a_x) = D(a_x^{n_x})$ we see that we may assume $a_x = h_x$.
Since X is quasi-compact (1.1.5(iii)) we can find a finite subcovering $X = \bigcup_{i=1}^{s} D(h_i)$ of the covering by the $D(h_x)$. There exist $g_i \in k[X]$ such that on $D(h_i)$ our function f equals $g_i h_i^{-1}$ $(1 \leq i \leq s)$. Since $g_i h_i^{-1}$ and $g_j h_j^{-1}$ coincide on $D(h_i) \cap D(h_j)$, whereas $h_i h_j$ vanishes outside that set, we have $h_i h_j (g_i h_j - g_j h_i) = 0$. Since the $D(h_i)$ cover X, the ideal generated by $h_1^2, \ldots, h_s^2$ is the ring $k[X]$. So there exist $a_i \in k[X]$ with

$$\sum_{i=1}^{s} a_i h_i^2 = 1.$$

Let $x \in U_j$. Then

$$h_j^2(x) \sum_{i=1}^{s} a_i(x) g_i(x) h_i(x) = \sum_{i=1}^{s} a_i(x) h_i^2(x) h_j(x) g_j(x) =$$

$$= h_j^2(x) f(x).$$

It follows that $f = \phi(\sum_{i=1}^{s} a_i g_i h_i)$, which proves the theorem.

1.4.6. <u>Exercise</u>. Let D(f) be a principal open subset of X. Show that there is an isomorphism onto $\mathcal{O}_X(D(f))$ of the algebra $k[X]_f = k[X][T]/(1-fT)$ ($k[X]_f$ is isomorphic to the ring of fractions $S^{-1}k[X]$, where $S = (f^h)_{h \geq 0}$, cf. 1.4.4(2)).

1.4.7. Morphisms.

Let $(X,\mathcal{O}_X)$ and $(Y,\mathcal{O}_Y)$ be two ringed spaces. Let $\phi: X \to Y$ be a continuous map. If f is a function on an open set $V \subset Y$, denote by $\phi_V^* f$ the composite function $f \circ \phi$ on the open set $\phi^{-1}V \subset X$. We say that ϕ is a morphism of ringed spaces if, for each open $V \subset Y$, we have that ϕ_V^* maps $\mathcal{O}_Y(V)$ into $\mathcal{O}_X(\phi^{-1}V)$. If, moreover, $(X,\mathcal{O}_X)$ and $(Y,\mathcal{O}_Y)$ are affine algebraic varieties, a morphism $\phi: X \to Y$ of ringed spaces is called a morphism of affine algebraic varieties.

If X is a subset of Y, ϕ the injection $X \to Y$ and $\mathcal{O}_X = \mathcal{O}_Y|X$ (see 1.4.2) then ϕ defines a morphism of ringed spaces in the previous sense. In that case we say that $(X,\mathcal{O}_X)$ is a ringed subspace of $(Y,\mathcal{O}_Y)$.

A morphism $\phi: X \to Y$ of affine varieties defines an algebra homomorphism $\phi^*: k[Y] \to k[X]$, as follows from 1.4.5. Conversely, if $\psi: k[Y] \to k[X]$ is an algebra homomorphism, it follows from the definitions that there exists a morphism $\tilde{\psi}: X \to Y$ with $(\tilde{\psi})^* = \psi$. Moreover, we have $(\psi^*)^\sim = \phi$. In fact, if $x \in X$ and $M_{X,x}$ is the corresponding maximal ideal of k[X] (see 1.3.3 (i)) then $\psi^{-1}M_{X,x}$ is a maximal ideal of k[Y], say $M_{Y,y}$. We then have $\tilde{\psi}(x) = y$.

1.4.8. Exercise. (1) Complete the proofs of these statements.

(2) Make affine k-algebras and affine algebraic varieties into categories and show that these two categories are anti-equivalent.

(3) A morphism of affine varieties $\phi: X \to Y$ is an isomorphism if and only if the algebra homomorphism $\phi^*: k[Y] \to k[X]$ is an isomorphism.

1.5. Products.

1.5.1. Let X and Y be two affine algebraic varieties. In accordance with the general notion of product in a category [26, Ch. I, §7] we say that a product of X and Y is an affine algebraic variety Z, together with morphisms $p: Z \to X$, $q: Z \to Y$ such that the following holds: for any triple (Z',p',q') of an affine variety Z' together with morphisms $p': Z' \to X$, $q': Z' \to Y$ there exists a unique morphism $r: Z' \to Z$ such that $p' = p \circ r$, $q' = q \circ r$. Put $k[X] = A$, $k[Y] = B$, $k[Z] = C$. Using 1.4.7 we see that C has the following property: there exist k-algebra homomorphisms $a: A \to C$, $b: B \to C$ such that for any triple (C',a',b') of an affine k-algebra and homomorphisms $a': A \to C'$, $b': B \to C'$ there is a unique k-algebra homomorphism $c: C \to C'$ with $a' = c \circ a$, $b' = c \circ b$.
Working in the category of k-algebras (i.e. forgetting the condition to be an affine k-algebra) it follows from familar properties of tensor products (see e.g. [26, Ch. XVI, §4]) that $C = A \otimes_k B$ and $a(x) = x \otimes 1$, $b(y) = 1 \otimes y$ satisfy our requirements.

1.5.2. Lemma. *Let A be a k-algebra of finite type and B any k-algebra. If A and B are reduced algebras (resp. integral domains) then the same holds for* $A \otimes_k B$.
Assume A and B reduced. Let $\sum_{i=1}^{n} a_i \otimes b_i$ be a nilpotent of $A \otimes_k B$. We may assume the b_i to be linearly independent over k. For any homomorphism $h: A \to k$ we have that $h \otimes \mathrm{id}$ is a homomorphism $A \otimes_k B \to B$. Consequently, $\sum_{i=1}^{n} h(a_i)b_i$ is a nilpotent element in B, hence 0. The linear independence of the

b_i shows that all $h(a_i)$ are 0, for all h. This means that all a_i are 0, proving that $A \otimes_k B$ is reduced. Next let A and B be integral domains. Let $x,y \in A \otimes_k B$, $xy = 0$. Write $x = \sum_i a_i \otimes b_i$, $y = \sum_j c_j \otimes d_j$, the sets $\{b_i\}$ and $\{d_j\}$ being linearly independent in B over k. An argument similar to the one just given then shows that $a_i c_j = 0$ for all i and j, from which it follows that x or y is 0.

1.5.3. Exercise. Show that 1.5.2 is true for any two k-algebras A,B.

1.5.4. Theorem. (i) *A product $X \times Y$ of two affine varieties X and Y exists, and is unique up to isomorphism;*
(ii) *If X and Y are irreducible, the same is true for $X \times Y$.*
By what we saw in 1.5.1 it suffices to show that if A and B are affine k-algebras (resp. affine k-algebras which are integral domains) the same is true for $A \otimes_k B$. This is a consequence of 1.5.2. The uniqueness statement of (i) follows from the definition of products, in a familiar manner.

1.5.5. Exercise. (1) Show that the set $X \times Y$ can be identified with the product set of X and Y.
(2) The Zariski topology on $X \times Y$ is finer then the product topology on $X \times Y$. Give an example where these topologies do not coincide.

1.6. Prevarieties and varieties.

1.6.1. Prevarieties.

A prevariety (over k) is a quasi-compact ringed space $(X, \mathcal{O}_X)$

(or simply X) such that any point has an open neighbourhood U with the property that the induced ringed space $(U, \mathcal{O}_X|U)$ is isomorphic to an affine algebraic variety (over k). Such a U is then called an affine open set of X.
Let X and Y be prevarieties. A morphism $\phi: X \to Y$ of prevarieties is by definition a morphism of the ringed spaces. A subprevariety of a prevariety is a ringed subspace (1.4.7) which is isomorphic to a prevariety.

1.6.2. Exercises. (1) A prevariety is a noetherian topological space (Hint: consider a descending chain of closed subsets).
(2) If X is an irreducible prevariety, and U an affine open set, then U is irreducible.

The notion of a product of prevarieties is defined in the same way as in 1.5.1.

1.6.3. Proposition. *A product of two prevarieties exists, and is unique up to isomorphism.*
Let X and Y be two prevarieties, and let $X = \bigcup_{i=1}^{m} U_i$, $Y = \bigcup_{j=1}^{n} V_j$ be finite coverings by affine open sets. The underlying set of $X \times Y$ will be the set theoretic product $X \times Y$, which is covered by the sets $U_i \times V_j$. On each of these we have a structure of affine variety (by 1.5.4 and 1.5.5(1)). We declare a set $U \subset X \times Y$ to be open if and only if its intersection with all $U_i \times V_j$ is open. This defines a topology on $X \times Y$. A function in an open neighbourhood of $x \in U_i \times V_j$ is regular in x if its restriction to $U_i \times V_j$ is regular in x, in the sense of the structure of affine variety defined on $U_i \times V_j$. This

defines a structure of ringed space on $X \times Y$. One verifies that $X \times Y$ has the required properties. The uniqueness is proved by a standard categorical argument.

1.6.4. Exercise. Fill in the details of the proof of 1.6.3.

1.6.5. Separation axiom.

Let X be a prevariety, let Δ_X be the diagonal in $X \times X$, i.e.

$$\Delta_X = \{(x,x) \mid x \in X\},$$

and let $i: X \to \Delta_X$ be the canonical map. We make Δ_X into a ringed subspace of $X \times X$ (see 1.4.7).

1.6.6. Example. Let X be an affine variety. Then Δ_X is a closed subset of $X \times X$, viz. the closed set $V_{X\times X}(I)$ defined by the ideal I in $k[X \times X]$ generated by the elements $f \otimes 1 - 1 \otimes f$, with $f \in k[X]$. From $k[X \times X]/I \simeq k[X]$ we see that i defines an isomorphism of ringed spaces $X \tilde{\to} \Delta_X$.

1.6.7. Exercise. Prove the assertions made in 1.6.6.

1.6.8. Lemma. $i: X \to \Delta_X$ _defines an isomorphism of ringed spaces, in all cases._

Cover Δ_X by open sets of the form $U \times U$, with U affine open in X, and apply the result of 1.6.6.

The prevariety X is said to be a _variety_ (or an algebraic variety over k, or a k-variety) if the following holds.

(_Separation axiom_): Δ_X _is closed in_ $X \times X$.

By 1.6.6 an affine variety is a variety. See 1.6.12(1) for an example of a prevariety which is not a variety.

A morphism $\phi: X \to Y$ of varieties is, by definition, a morphism of the underlying prevarieties.

1.6.9. Exercises. (1) Show that a topological space X is a Hausdorff space if and only if the diagonal Δ_X is closed in $X \times X$, provided with the product topology.
(2) If X and Y are varieties, then so is $X \times Y$.
(3) Give a definition of the notion of sub-prevariety of a prevariety. Show that a sub-prevariety of a variety is a variety.
(4) Let X be a variety. Define a induced structure of variety on an open or closed subset of X.

One needs the separation axiom to establish results like the following.

1.6.10. *Proposition*. *Let X be a variety, let Y be any prevariety.*
(i) If $\phi: Y \to X$ is a morphism, its graph $\Gamma_\phi = \{(y,\phi(y)) | y \in Y\}$ is closed in $Y \times X$;
(ii) If $\phi,\psi: Y \to X$ are two morphisms which coincide on a dense set, then $\phi = \psi$.
To prove (i) consider the continuous map $Y \times X \to X \times X$ sending (y,x) to $(\phi(y),x)$. Then Γ_ϕ is the inverse image of Δ_X, so is closed. This proves (i). In the situation of (ii) one sees, similarly, that $\{y | \phi(y) = \psi(y)\}$ is closed. This implies (ii).

The following result contains a criterion for prevarieties to be varieties.

1.6.11. Proposition. (i) *Let X be a variety, let U and V be affine open sets in X. Then U ∩ V is an affine open set and the images under restriction maps of* $\mathcal{O}_X(U)$ *and* $\mathcal{O}_X(V)$ *in* $\mathcal{O}_X(U \cap V)$ *generate this algebra;*

(ii) *Let X be a prevariety, and let* $X = \bigcup_{i=1}^{m} U_i$ *be a covering by affine open sets. Then X is a variety if and only if the following condition is satisfied: for each pair (i,j) the intersection* $U_i \cap U_j$ *is an affine open set and the images under restriction maps of* $\mathcal{O}_X(U_i)$ *and* $\mathcal{O}_X(U_j)$ *in* $\mathcal{O}_X(U_i \cap U_j)$ *generate this algebra.*

In the situation of (i) we have that $\Delta_X \cap (U \times V)$ is closed in $U \times V$. Now i induces an isomorphism of ringed spaces $U \cap V \tilde{\to} \Delta_X \cap (U \times V)$. It follows that $U \cap V$ is affine and that regular functions on $\Delta_X \cap (U \times V)$ are restrictions of regular functions on $U \times V$, i.e. of elements of $k[U] \otimes_k k[V]$. This proves (i).

That the condition of (ii) is necessary follows from (i). If it is satisfied then $\Delta_X(U_i \times U_j)$, being isomorphic to $U_i \cap U_j$, is an affine algebraic variety whose coordinate algebra is a quotient of $\mathcal{O}_{X\times X}(U_i \times U_j)$. This implies that $\Delta_X \cap (U_i \times U_j)$ is closed in $U_i \times U_j$, and Δ_X is closed in $X \times X$.

1.6.12. Exercises. (1) Define the "line with the point 0 doubled" X as follows. As a set, X is the union of $\mathbb{A}^1$ and a point {0'}, moreover $\mathbb{A}^1$ is an affine open set, with its usual structure of variety. Define $\phi: \mathbb{A}^1 \to X$ by: $\phi(x) = x \in \mathbb{A}^1$ if $x \neq 0$, $\phi(0) = 0'$. Define $\phi\mathbb{A}^1$ to be affine open, with the transported structure of variety.

Show that X is a prevariety which is not a variety.

(2) Define the projective line $\mathbb{P}^1$ in a similar manner: $\mathbb{P}^1 = \mathbb{A}^1 \cup \{\infty\}$, but now ϕ is defined by $\phi(x) = x^{-1}$ $(x \neq 0)$, $\phi(0) = \infty$.

Show that $\mathbb{P}^1$ is a variety. Show that $\mathcal{O}_{\mathbb{P}^1}(\mathbb{P}^1) = k$ and deduce that any morphism of varieties ϕ: $\mathbb{P}^1 \to X$ with X affine, is constant.

(3) Let $U \subset A^n$ be open and non-empty. If $f \in \mathcal{O}_{\mathbb{A}^n}(U)$, there exist $g,h \in k[T_1,\ldots,T_n]$, without nontrivial common factor, such that h does not vanish on U and that $f(x) = g(x)h(x)^{-1}$ for $x \in U$. (Hint: cover U by sets D(a) and use 1.4.6).

(4) Let $X = \mathbb{A}^n - \{0\}$ with the induced structure of variety. Show that X is not an affine variety if $n > 2$ (use the previous exercise).

1.7. Projective space.

The most important examples of non-affine varieties (and practically the only ones which we shall encounter) are the projective spaces and their closed subvarieties.

1.7.1. $\mathbb{P}^n$.

The underlying set of projective n-space $\mathbb{P}^n$ is the set of all 1-dimensional subspaces of k^{n+1}, or, equivalently, $k^{n+1}-\{0\}$ modulo the equivalence relation: $x \sim y$ if and only there is $a \in k^* = k - \{0\}$ such that $y = ax$. We write x^* for the equivalence class of x. If $x = (x_0,\ldots,x_n)$, we call the x_i homogeneous coordinates of x^*.

For $0 \leqslant i \leqslant n$, put

$$U_i = \{(x_0,\dots,x_n)^* \in \mathbb{P}^n \mid x_i \neq 0\}.$$

We declare these sets to be open in $\mathbb{P}^n$. Define a bijection $\phi_i: U_i \to \mathbb{A}^n$ by

$$\phi_i(x_0,\dots,x_n)^* = (x_i^{-1}x_0,\dots,x_i^{-1}x_{i-1},x_i^{-1}x_{i+1},\dots,x_i^{-1}x_n),$$

and transport the structure of affine variety of $\mathbb{A}^n$ to U_i, via ϕ_i. Then $\phi_i(U_i \cap U_j)$ is an open set $D(f)$ of $\mathbb{A}^n$. In fact, identifying $k[\mathbb{A}^n]$ with $k[T_1,\dots,T_n]$, we take $f = T_j$ if $j < i$, $f = 1$ if $j = i$ and $f = T_{j-1}$ if $j > i$.

We define a topology on $\mathbb{P}^n$ by defining $U \subset X$ to be open if all $U \cap U_i$ are open in U_i $(0 \leqslant i \leqslant n)$. Moreover, since $\mathcal{O}_{U_i}|U_i \cap U_j$ and $\mathcal{O}_{U_j}|U_i \cap U_j$ coincide, it makes sense to define a <u>sheaf</u> $\mathcal{O}_X$ <u>of</u> <u>regular</u> <u>functions</u> as follows. A function f defined in a neighbourhood of $x \in \mathbb{P}^n$ is regular in x if f is regular in x as a function defined locally in U_i, for some i with $x \in U_i$. Then $\mathcal{O}(U)$ is the algebra of functions in U, regular in all $x \in U$.

We now have a structure of prevariety on $\mathbb{P}^n$. This is, in fact, a variety. For from the definitions we see that $\mathcal{O}_{\mathbb{P}^n}(U_i)$ is the set of functions on U_i, whose value in $(x_o,\dots,x_n)^*$ is a polynomial in $x_i^{-1}x_0,\dots,x_i^{-1}x_n$. Similarly, $\mathcal{O}_{\mathbb{P}^n}(U_i \cap U_j)$ is the set of functions which are polynomial in $x_i^{-1}x_0,\dots,x_i^{-1}x_n,x_j^{-1}x_0,\dots,x_j^{-1}x_n$. It is then clear that the condition of 1.6.11(ii) is satisfied.

$\mathbb{P}^n$, with the variety structure just defined, is <u>projective</u> n-<u>space</u>. If $n = 1$ we recover the variety of 1.6.12(2) (check this). A <u>projective variety</u> is a closed subset of $\mathbb{P}^n$, together with its induced structure of variety. A <u>quasi-projective variety</u> is an open subvariety of a projective one.

1.7.2. <u>Exercises</u>. (1) An invertible linear map of k^{n+1} induces an isomorphism of $\mathbb{P}^n$.

(2) Let V be a finite dimensional vector space over k. Define a variety $\mathbb{P}(V)$ whose underlying point set is the set of 1-dimensional subspaces of V, and which is isomorphic to $\mathbb{P}^{n-1}$ ($n = \dim V$).

1.7.3. <u>Closed sets in</u> $\mathbb{P}^n$.

We shall give another description of closed sets of $\mathbb{P}^n$. Let $S = k[T_0,\dots,T_n]$ be the polynomial algebra in n+1 indeterminates. An ideal I in S is <u>homogeneous</u> if it is generated by homogeneous polynomials, or, equivalently if $f = f_0+\dots+f_k \in I$ (where f_i is homogeneous of degree i) implies that $f_i \in I$. Let $S_+ \subset S$ be the homogeneous ideal of all polynomials with constant term 0.

If I is a proper homogeneous ideal in S, then if $x \in k^{n+1}$ is a zero of I, the same is true for all ax, $a \in k$.

Hence we can define a set $V^*(I) \subset \mathbb{P}^n$ by

$$V^*(I) = \{x^* \in \mathbb{P}^n \mid x \in V_{k^{n+1}}(I)\}.$$

1.7.4. <u>Proposition</u>. <u>The closed sets in</u> $\mathbb{P}^n$ <u>coincide with the sets</u> $V^*(I)$, I <u>a proper homogeneous ideal in</u> S.

It is easy to see that $V^*(I) \cap U_i$ is closed for all i, so $V^*(I)$ is closed. To prove 1.7.4 it is now sufficient to show that any open set U is the complement of some $V^*(I)$. Using an analogue of 1.1.3(d) we see that it suffices to do this if $U = \phi_i^{-1}(D(f)) \subset U_i$, where $f \in k[T_1,\ldots,T_n]$. Now there is a homogeneous polynomial $f^* \in k[T_0,\ldots,T_n]$, divisible by T_i, such that $f^*(x_1,\ldots,x_{i-1},1,x_{i+1},\ldots,x_n) = f(x_1,\ldots,x_n)$. It is clear that U is the complement of $V^*(f^*S)$.

1.7.5. Exercises. (1) Let $f \in S$ be homogeneous. Let S_f^* be the algebra of quotients gf^{-h}, where $g \in S$ is homogeneous and $\deg g = h \deg f$.
Show that $D^*(f) = \mathbb{P}^n - V^*(fS)$ is an affine open subset of $\mathbb{P}^n$ and that $\mathcal{O}_{\mathbb{P}^n}(D^*(f)) \simeq S_f^*$.
(2) Let $V^*(I)$ be as in 1.7.3. Show that $V^*(I) = \phi$ if and only if there exists $N > 0$ such that $T_i^N \in I$ $(0 \leqslant i \leqslant n)$. Show that $V^*(I)$ is irreducible if and only if I is a prime ideal.

1.7.6. Products of projective spaces.
For the next proposition we need an imbedding of $\mathbb{P}^m \times \mathbb{P}^n$ in $\mathbb{P}^{mn+m+n}$. We view the last space as being described by homogeneous coordinates $(x_{ij})_{0 \leqslant i \leqslant m, 0 \leqslant j \leqslant n}$. Define a map $\phi: \mathbb{P}^m \times \mathbb{P}^n \to \mathbb{P}^{mn+m+n}$ by

$$\phi((x_0,\ldots,x_m)^*,(y_0,\ldots,y_n)^*) = (x_i y_j)^*_{0 \leqslant i \leqslant m, 0 \leqslant j \leqslant n}.$$

1.7.7. Proposition. (i) *The image of ϕ is a closed subset $V^{m,n}$ of $\mathbb{P}^{mn+m+n}$;*
(ii) *ϕ defines an isomorphism of varieties $\mathbb{P}^m \times \mathbb{P}^n \tilde{\to} V^{m,n}$.*
We have that $V^{m,n}$ is the closed set defined by the homogeneous ideal in $k[T_{ij}]$ generated by the elements $T_{ij}T_{pq} - T_{pj}T_{iq}$. This establishes (i), and (ii) is a matter of verification. We leave the details to the reader.

1.7.8. Corollary. *The product of two projective varieties is again a projective variety.*

1.7.9. Exercise. Describe $\mathbb{P}^1 \times \mathbb{P}^1$, as a subvariety $V^{1,1}$ of $\mathbb{P}^3$.

We next discuss some facts needed in the next chapter.

1.8. Dimension.

1.8.1. Let X be an irreducible variety. If X is affine and U is a principal open subset D(f) (1.3.5) then $k[U] = k[X]_f$ for some $f \in k[X]$ (1.4.6) and it follows that the quotient fields of the integral domains k[X] and k[U] can be identified. Using 1.3.6(ii) this implies the same fact for any affine open subset of the irreducible affine variety X. Now let X be arbitrary. Then 1.6.11(i) and the preceding remarks imply that for any two affine open sets U,V the quotient fields of k[U] and k[V] can be canonically identified. It follows that we can speak of the *quotient field* k(X) of X. Its transcendence degree over k (see [26 , Ch. X, §1]) is the *dimension* of X, denoted dim X. If X is reducible and $(X_i)_{1 \leq i \leq n}$ are its components, then we define

$$\dim X = \max_{1 \leq i \leq n} (\dim X_i).$$

If X is affine and $k[X] = k[x_1, \ldots, x_r]$ then $k(X) = k(x_1, \ldots, x_r)$, and dim X is the maximal number of algebraically independent elements among the x_i.

1.8.2. Proposition. *If X is irreducible and Y is a proper*

<u>irreducible closed subvariety of</u> X, <u>then</u> dim Y < dim X.

It suffices to consider the case that X is affine. Let k[X]=A, then k[Y] = A/P, where P is a nonzero prime ideal. Let $A = k[x_1,\ldots,x_r]$ and let y_i be the image of x_i in A/P. Let d = dim X, e = dim Y. Then we may assume that $y_1,\ldots,y_e$ are algebraically independent over k, and clearly the same is true for $x_1,\ldots,x_e$, whence $e \leq d$. If equality holds, take any nonzero $f \in P$. There is a relation $F(f,x_1,\ldots,x_d) = 0$, where $F \in k[T_0,T_1,\ldots,T_d]$, and we may assume that F is not divisible by T_0. But then we have a nontrivial relation $F(0,y_1,\ldots,y_e) = 0$, a contradiction. Hence dim Y < dim X.

1.8.3. <u>Proposition</u>. <u>Let</u> X an Y <u>be irreducible varieties</u>. <u>Then</u> dim X × Y = dim X + dim Y.

We may assume X and Y to be affine. Let $k[X] = k[x_1,\ldots,x_r]$, $k[Y] = k[y_1,\ldots,y_s]$, d = dim X, e = dim X. Assume that $\{x_1,\ldots,x_d\}$ and $\{y_1,\ldots,y_e\}$ are maximal algebraically independent sets, respectively. Then $\{x_1 \otimes 1,\ldots x_d \otimes 1, 1 \otimes y_1,\ldots,1 \otimes y_e\}$ is a maximal set of algebraically independent elements in $(x_i \otimes 1, 1 \otimes y_j)_{1 \leq i \leq r,\ 1 \leq j \leq s}$.

This proves the proposition.

1.8.4. <u>Exercises</u>. (1) $\dim \mathbb{A}^n = n$, $\dim \mathbb{P}^n = n$.

(2) A 0-dimensional variety consists of finitely many points.

(3) Let $f \in k[T_1,\ldots,T_n]$ be irreducible. The set of its zeros is an (n-1)-dimensional affine irreducible subvariety of $\mathbb{A}^n$.

1.9. Some results on morphisms.

1.9.1. Lemma. Let X and Y be affine varieties, let ϕ: X → Y be a morphism and ϕ^*: k[Y] → k[X] the associated algebra homomorphism.

(i) If ϕ^* is surjective then ϕ maps X onto a closed subset of Y;

(ii) ϕ^* is injective if and only if $\overline{\phi X} = Y$ (the bar denoting closure);

(iii) If X is irreducible then so is $\overline{\phi X}$, and $\dim \overline{\phi X} \leqslant \dim X$.

Let I = Ker ϕ^*. If ϕ^* is surjective then $\phi X = V_Y(I)$, whence (i). Also, ϕ^* is injective if and only if $I_Y(\phi X) = 0$, whence (ii). The first point of (iii) follows from 1.2.3, and the last point from (ii) (applied to the restriction morphism $X \to \overline{\phi X}$.

The image ϕX need not be closed. As an example, let $X = \{(x,y) \in \mathbb{A}^2 | xy = 1\}$, $Y = \mathbb{A}^1$ and define $\phi(x,y) = x$. Then ϕX is the open set $\mathbb{A}^1 - \{0\}$ and $\overline{\phi X} = Y$.

1.9.2. We shall establish some results about the images under morphisms. For this we need some algebraic results, about the following situation.

Let B be a ring and A a subring such that B is of finite type over A. We want to extend a homomorphism of A into an algebraically closed field to a homomorphism of B. We first establish a result for a special case, viz. that B = A[b], for some b ∈ B. Then B ≃ A[T]/I, where I is the ideal of the f ∈ A[T] with f(b) = 0. We denote by J(I) ⊂ A the set of leading coefficients of the polynomials of I. It is obvious that $J(I)$ is an ideal in A.

1.9.3. *Lemma. Assume moreover that* B *is an integral domain. Let* ϕ *be a homomorphism of* A *into the algebraically closed field* F *such that* $\phi J(I) \neq 0$. *Then* ϕ *can be extended to a homomorphism of* B *into* F.

Let $f = f_0 + f_1T + \dots + f_mT^m \in I$ and $\phi f_m \neq 0$. We may assume that the degree m of f is minimal, subject to these conditions. Then $m \geq 1$. We use induction on m. First extend ϕ to a homomorphism $A[T] \to F[T]$ in the obvious way. If the ideal $\phi I \subset \phi A[T]$ does not contain nonzero constants, it generates a proper ideal in $F[T]$. Let $z \in F$ be a zero of that ideal. It is then immediate that $\phi b = z$ defines an extension of ϕ to B.

Now assume that ϕI does contain a nonzero constant. This means that I contains a polynomial $g = g_0 + g_1T + \dots + g_nT^n$ with $\phi(g_0) \neq 0$, $\phi(g_i) = 0$ $(i > 0)$. The division algorithm shows that there exist $q, r \in A[T]$ and an integer $d \geq 0$ such that $a_m^d g = qf + r$, and that $\deg r < m$. Then $\phi(a_m)^d \phi(g_0) = \phi(q).\phi f + \phi r$. Since ϕf has degree $m > 0$, we have that ϕr is also a nonzero constant. This means that we may assume $n < m$. If $m = 1$ this shows that g cannot exist, proving the assertion in that case. So assume $m > 1$ and that the assertion is true for smaller values.

If $h = h_0 + h_1T + \dots + h_sT^s$ and $h_s \neq 0$, put $\tilde{h} = h_s + h_{s-1}T + \dots + h_0T^s$. It is easy to see that the polynomials $T^a\tilde{h}$, with $a \geq 0$, $h \in I$ span an ideal $\tilde{I}$ in $A[T]$. Moreover $\tilde{I}$ contains a nonzero constant if and only if I contains a nonzero polynomial aT^s, which can only be if $b = 0$, $A = B$ (since B is an integral domain).

If $\tilde{I}$ does not contain nonzero constants, put $\tilde{B} = A[T]/\tilde{I} = A[\tilde{b}]$. Then $\tilde{I}$ contains the polynomial $\tilde{g}$ of degree $< m$. By induction we may assume that ϕ has been extended to a homomorphism $\tilde{B} \to F$. This means that there is $z \in F$ such that for all $h = h_0 + h_1 T + \dots + h_s T^s$ in I we have

$$\phi(h_s) + \phi(h_{s-1}) z + \dots + \phi(h_0) z^s = 0.$$

Taking $h = g$ we see that $z = 0$. Taking $h = f$ we then obtain the contradiction $\phi(f_m) = 0$. This implies that g cannot exist, and the lemma follows.

1.9.4. *Lemma. Let* B *be an integral domain and* A *a subring such that* B *is of finite type over* A. *Given* $b \neq 0$ *in* B *there exists* $a \neq 0$ *in* A *such that any homomorphism* ϕ *of* A *into the algebraically closed field* F *with* $\phi(a) \neq 0$ *can be extended to a homomorphism* $\phi: B \to F$ *with* $\phi(b) \neq 0$.

We have $B = A[b_1,\dots,b_n]$. By induction we may assume $n = 1$, i.e. $B = A[b_1] \simeq A[T]/I$, as before. First assume $I \neq 0$. Let $f \in I$ be nonzero and of minimal degree, let a' be its leading coefficient. The division algorithm shows that $g \in I$ if and only if for some $d \geq 0$ we have that $(a')^d g$ is divisible by f. Now let $h \in A[T]$ represent the given element b. Since $b \neq 0$ we have $h \notin I$. Since f is irreducible over the quotient field of A, it follows that f and h are coprime over that field. Hence there exist $u,v \in A[T]$ and $a'' \in A - \{0\}$, such that

$$uh + vf = a''.$$

It now follows that a = a'a" satisfies our requirements. For if $\phi a \neq 0$, we have that ϕ can be extended to B, by the previous lemma. Then $\phi(u(b_1))\phi(b) = \phi(a'') \neq 0$, whence $\phi(b) \neq 0$. This settles the case that $I \neq 0$. If I = 0 the argument is quite easy and is left to the reader.

1.9.5. Theorem. Let ϕ: X → Y be a morphism of varieties. Then ϕX contains a non-empty open piece of $\overline{\phi X}$.
Using suitable coverings of X and Y by affine open sets one sees that it suffices to consider the case that X and Y are affine. If $X_1, \ldots, X_r$ are the components of X, then $\overline{\phi X} = \overline{\phi X_1} \cup \ldots \cup \overline{\phi X_r}$ and one sees that it suffices to assume X irreducible. We may then also replace Y by $\overline{\phi X}$. After these reductions we are reduced to the case that X and Y are affine and irreducible and that $\overline{\phi X} = Y$. In that case (using 1.9.1 (ii)) we see that the theorem is a consequence of 1.9.4.

1.9.6. Exercises. (1) In 1.9.3 the assumption that B be an integral domain can be dropped.
(2) Let X be a variety. A subset of X is locally closed if it is the intersection of an open and a closed subset; a subset which is a union of finitely many locally closed subsets is constructible.
Let ϕ: X → Y be a morphism. Show that ϕX is constructible. (Hint: use induction on dim Y, using 1.8.2 and 1.9.4). Show that the image of a constructible subset of X is constructible.

Appendix.

1.9.4, the proof of which is independent of the rest of this chapter, can be used to prove the Nullstellensatz 1.1.2.
To prove 1.1.2(i), let $M \supset I$ be a maximal ideal (which exists, for example because S is noetherian). Then S/M is a field, which is also a k-algebra of finite type, and 1.1.2(i) follows from the following, more general, result:
Let k be any field, let l be an extension field of k which is also a k-algebra of finite type. Then l is a finite extension of k.
Apply 1.9.4 with A = k, B = l, b = 1, F an algebraic closure of k. It follows that l is isomorphic to a subfield of F, i.e. is algebraic over k. This implies the result.
To prove 1.1.2(ii) we use that a radical ideal in S is an intersection of prime ideals (this follows from the noetherian property, by an argument similar to that of 1.2.4(i)). Now let $f \notin \sqrt{I}$, and let $P \supset I$ be a prime ideal not containing f. Apply 1.9.4 with A = k, B = A/P, b = f mod P, F = k. It follows that there is a zero of P which is not a zero of f, whence 1.1.2(ii).

Notes.

This first chapter contains mostly standard elementary material from algebraic geometry, and needs few comments.
For more about algebraic geometry refer to Hartshorne's book [21].
Since sheaves are not used in the sequel, we have not spoken

in 1.4 of the general notion of sheaf, but only of sheaves of functions. Generalities about sheaves can be found in [18].

In the literature, 1.9.4 is usually proved by using valuations. This is not really necessary. We have used here an elementary approach, which goes back to Chevalley and Weil (see [36 , p.30-31]). The result 1.9.3 will also be used in chapter 4 (proof of 4.2.3).

2. Linear algebraic groups, first properties.

k is an algebraically closed field.

2.1. Algebraic groups.

2.1.1. An *algebraic group* over k is a variety G, which is also a group, such that the maps defining the group structure π: $G \times G \to G$ with $\pi(x,y) = xy$ and ι: $G \to G$ with $x = x^{-1}$, are morphisms of varieties. Recall that by 1.5.5(1) we may view the set of points of the variety $G \times G$ to be the product set. If the underlying variety is affine, then G is a *linear algebraic group*. These are the ones we shall be concerned with (it is usual to use the adjective "linear" instead of "affine").

Let G be a linear algebraic group, and put $A = k[G]$. Because of the facts stated in 1.4.7, the group structure of G is defined by algebra homomorphisms $\pi^*: A \to A \otimes_k A$, $\iota^*: A \to A$, and the identity element e is a homomorphism $A \to k$. These have to satisfy a number of axioms, reflecting the group axioms (see the next exercise). Let G and G' be algebraic groups. A morphism of varieties ϕ: $G \to G'$ is a *homomorphism of algebraic groups* if it is also a group homomorphism. ϕ is an *isomorphism of algebraic groups* if it is an isomorphism of varieties and of groups. Automorphisms are defined similarly.

By 1.5.5(1) we can view the underlying set of the product variety $G \times G'$ to be the set-theoretical product. We can then provide $G \times G'$ with the usual direct product group structure. This makes $G \times G'$ into an algebraic group, the *direct product* of the algebraic groups G and G'.

2.1.2. Exercises.

(1) Let A, π^*, ι^* and e be as above. Denote by $\gamma: G \to G$ the constant morphism which maps everything into e. The corresponding algebra homomorphism $\gamma^*: A \to A$ is given by $\gamma^* f = f(e)$. Show that the group axioms for G are expressed by the commutativity of the following diagrams:

$$\begin{array}{ccc} A \otimes A \otimes A & \xleftarrow{\ \pi^* \otimes \mathrm{id}\ } & A \otimes A \\ {\scriptstyle \mathrm{id} \otimes \pi^*} \uparrow & & \uparrow {\scriptstyle \pi^*} \\ A \otimes A & \xleftarrow{\ \pi^*\ } & A \end{array} \qquad \begin{array}{ccc} A & \xleftarrow{\ \iota^* \otimes \mathrm{id}\ } & A \otimes A \\ {\scriptstyle \mathrm{id} \otimes \iota^*} \uparrow & \nwarrow & \uparrow {\scriptstyle \pi^*} \\ A \otimes A & \xleftarrow{\ \pi^*\ } & A \end{array}$$

$$\begin{array}{ccc} A & \xleftarrow{\ \gamma^* \otimes \mathrm{id}\ } & A \otimes A \\ {\scriptstyle \mathrm{id} \otimes \gamma^*} \uparrow & \overset{\mathrm{id}}{\nwarrow} & \uparrow {\scriptstyle \pi^*} \\ A \otimes A & \xleftarrow{\ \pi^*\ } & A \end{array}$$

(2) Check that $G \times G'$, as defined above, is indeed an algebraic group.

(3) Let G be an algebraic group and H a closed subset of G which is also a subgroup. Let $\phi: H \to G$ be the inclusion map. There is a structure of algebraic group on H such that ϕ is a homomorphism of algebraic groups.

(4) Define a notion of the pre-algebraic group, starting with a prevariety G, instead of a variety. Show that a pre-algebraic group is an algebraic group.

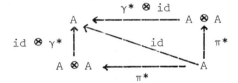

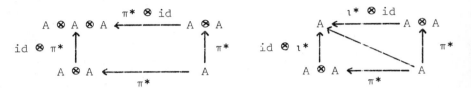

2.1.3. Examples.

(1) $G = \mathbb{A}^1$, with addition as group operation. This defines a linear algebraic group, with $A = k[T]$. Now $\pi^*: k[T] \to k[T] \otimes_k k[T] \simeq k[T,U]$ is given by $\pi^*T = T+U$, and $\iota^*: k[T] \to k[T]$ by $\iota^*T = -T$. The identity element e is the k-homomorphism $k[T] \to k$ sending T to 0. We denote this algebraic group by $\mathbb{G}_a$: it is the additive group.

(2) $G = \mathbb{A}^1-\{0\}$, with multiplication as group operation. We have $A = k[T,T^{-1}]$, and $\pi^*: k[T,T^{-1}] \to k[T,T^{-1}] \otimes k[T,T^{-1}] \simeq k[T,T^{-1},U,U^{-1}]$ is given by $\pi^*T = TU$. Moreover, $\iota^*T = T^{-1}$ and $e: k[T,T^{-1}] \to k$ sends T to 1.

This algebraic group is denoted by $\mathbb{G}_m$ or $\mathbb{GL}_1$: it is the multiplicative group. If n is any nonzero integer, then $\phi x = x^n$ defines a homomorphism of algebraic groups $\mathbb{G}_m \to \mathbb{G}_m$. If k has characteristic $p > 0$, and n is a power of p, then ϕ is an isomorphism of abstract groups but not of algebraic goups (since $\phi^*: A \to A$ is not surjective, see 1.4.8(3)).

(3) View the space $\mathbb{M}_n$ of all n×n-matrices as k^{n^2}. If $x \in \mathbb{M}_n$, let D(X) be its determinant. The general linear group $\mathbb{GL}_n$ is the principal open set $\{X \in \mathbb{M}_n | D(X) \neq 0\}$, with matrix multiplication as group operation.

We have $k[\mathbb{GL}_n] = k[T_{ij}, D^{-1}]_{1 \leq i,j \leq n}$, where $D = \det(T_{ij})$. In this case π^* is given by

$$\pi^*T_{ij} = \sum_{h=1}^{n} T_{ih} \otimes T_{hj},$$

and ι^*T_{ij} is the (i,j)-entry of the matrix $(T_{ij})^{-1}$. The identity e sends T_{ij} to δ_{ij}.

Since $\mathbb{M}_n$ is an irreducible variety, so is $\mathbb{GL}_n$. Its dimension is n^2.

(4) Any subgroup of $\mathbb{GL}_n$ which is closed in the Zariski topology of $\mathbb{GL}_n$, is a linear algebraic group. Here are a number of examples.

(a) a finite subgroup;

(b) $\mathbb{D}_n$, the group of nonsingular diagonal matrices;

(c) $\mathbb{T}_n$, the group of upper triangular matrices, defined by $\mathbb{T}_n = \{X = (x_{ij}) \in \mathbb{GL}_n \mid x_{ij} = 0 \text{ if } i > j\}$;

(d) $\mathbb{U}_n$, the group of unipotent upper triangular matrices, defined by $\mathbb{U}_n = \{X = (x_{ij}) \in \mathbb{GL}_n \mid x_{ij}=0 \text{ if } i > j,\ x_{ii} = 1\}$;

(e) The special linear group $\mathbb{SL}_n = \{X \in \mathbb{GL}_n \mid D(X) = 1\}$;

(f) The orthogonal group $\mathbb{O}_n = \{X \in \mathbb{GL}_n \mid X.{}^tX = 1\}$, where ${}^tX$ denotes the transpose of X;

(g) The special orthogonal group $\mathbb{SO}_n = \mathbb{O}_n \cap \mathbb{SL}_n$;

(h) The symplectic group $\mathbb{S}_{p2n}$ =

$$\left\{X \in \mathbb{GL}_{2n} \mid {}^tX \begin{pmatrix} 0 & 1_n \\ -1_n & 0 \end{pmatrix} X = \begin{pmatrix} 0 & 1_n \\ -1_n & 0 \end{pmatrix}\right\}.$$

(5) As examples of non-linear algebraic groups (not needed in the sequel) we mention the <u>elliptic curves</u>. These are closed subsets of the projective plane $\mathbb{P}^2$. Assuming for convenience that char k $\neq 2,3$ such a group G can be defined as the set of all $(x_0,x_1,x_2)^* \in \mathbb{P}^2$ (notations of 1.7.1) such that $x_0x_2^2 = x_1^3 + ax_1x_0^2 + bx_0^3$, where $a,b \in k$ are such that the polynomial $T^3 + aT + b$ has no multiple roots.

The neutral element e is $(0,0,1)^*$. The group operation of G is abelian, and is written as addition. It is such that if three distinct points of G are collinear (i.e. if their homogeneous coordinates satisfy a nontrivial linear equation

$c_0x_0 + c_1x_1 + c_2x_2 = 0$), then their sum is e. This defines the addition, and it is easy to check that if $x = (x_0, x_1, x_2)^* \in G$, then $-x = (x_0, x_1, -x_2)^*$.
The addition may, in fact, be given explicitly by formulas, which are, however, not very enlightening. A proof of the associativity of the group operation starting from these formulas would be quite clumsy.
These are better, more geometric, ways to deal with the group structure on such a curve. We refer to [21, p. 321].

2.1.4. Exercises.
(1) Let V be a finite dimensional vector space over k. Define a linear algebraic group GL(V) whose underlying abstract group is the group of all invertible linear maps of V, and which is isomorphic to $\mathbb{GL}_n$ (n = dim V).
(2) Check that all subgroups of $\mathbb{GL}_n$ of 2.1.3(4) are indeed closed.
(3) We have $A = k[\mathbb{SL}_2] = k[T_1, T_2, T_3, T_4]/(T_1T_4 - T_2T_3 - 1) = k[t_1, t_2, t_3, t_4]$ (t_i denoting the image of T_i). Let B be the subalgebra of A generated by the products t_it_j ($1 \leqslant i, j \leqslant 4$).
(a) Let π^* and ι^* define the group structure of $\mathbb{SL}_2$. Show that $\pi^*B \subset B \otimes_k B$, $\iota^*B \subset B$ and deduce that there is a algebraic group $\mathbb{PSL}_2$, whose coordinate algebra is B, together with a homomorphism of algebraic groups $\mathbb{SL}_2 \to \mathbb{PSL}_2$, whose kernel is $\{\pm 1\}$. It is not an isomorphism of algebraic groups;
(b) Assume char $k \neq 2$. Then B is the algebra of functions $f \in A$ such that $f(-X) = f(X)$, for all $X \in \mathbb{SL}_2$;
(c) Assume char $k = 2$. Then the homomorphism $\mathbb{SL}_2 \to \mathbb{PSL}_2$ of

(a) is an isomorphism of abstract groups.

(4) Show that the group $\mathbb{U}_n$ of 2.1.3(4d) is solvable (it is even nilpotent).

2.2. Elementary properties of algebraic groups.

Let G be an algebraic group. If $g \in G$, the maps $x \mapsto gx$ and $x \mapsto xg$ are isomorphisms of the variety G. We shall often use this.

2.2.1. Propostion. (i) *There is a unique irreducible component* G^0 *of* G *containing the identity element* e; *it is a closed normal subgroup of finite index*;

(ii) G^0 *is the connected component of* e, *for the Zariski topology*;

(iii) *Any closed subgroup* H *of* G *of finite index contains* G^0.

Let X and Y be irreducible components of G containing e. If π and ι are as in 2.1.1, it follows from 1.2.3 that $XY = \pi(X \times Y)$ is irreducible and that the same is true for its closure $\overline{XY}$. Since $X \subset \overline{XY}$, $Y \subset \overline{XY}$, it follows from 1.2.5(1) that we must have $X = \overline{XY}$, $Y = \overline{XY}$. Hence $X = Y$. We also see that $X.X = X$. Since ι is a homeomorphism of the topological space G, we see that X^{-1} is a component of G containing e, so $X^{-1} = X$. It follows that X is a closed subgroup. Similarly, for any $x \in G$ we have $xXx^{-1} = X$, and X is a normal subgroup. The cosets xX must be components of G. The last assertion of (i) now follows from 1.2.4. We have $G^0 = X$.

It follows from (i) that the complement of G^0 in G, being a finite union of components, is also closed. So G^0 is open and closed, and also connected (because irreducible). So G^0 must be the maximal connected subset of G containing e, whence (ii). This argument also shows that a subgroup H of (iii) is both open and closed. Since it contains e, we must have $H \subset G^0$, which proves (iii).

It follows from 2.2.1 that for an algebraic group the notions of connectedness and irreducibility coincide. It is usual to speak of a connected algebraic group, and not of an irreducible one. We follow this convention.
In the sequel, we always denote by G^0 the identity component of an algebraic group G. Dimensions being defined as in 1.8.1, it follows from 2.2.1 that $\dim G = \dim G^0$.

2.2.2. Exercises. (1) The groups $\mathbb{G}_a, \mathbb{G}_m, \mathbb{D}_n, \mathbb{T}_n, \mathbb{U}_n, \mathbb{GL}_n, \mathbb{SL}_n$ of 2.1.3 are all connected. If char $k \neq 2$, the group $\mathbb{O}_n$ is not connected (then $\mathbb{SO}_n$ and $\mathbb{S}p_{2n}$ are also connected, but this requires a bit more work, see 6.14).
(2) The variety $V(TU)$ in $\mathbb{A}^2$ is connected, but reducible (hence it cannot be the underlying variety of an algebraic group).
(3) Let G be a connected algebraic group, let N be a finite normal subgroup of G. Then N lies in the center of G (Hint: let $n \in N$ and consider the morphism $g \mapsto gng^{-1}$ of G into N).

2.2.3. Lemma. Let U and V be dense open subsets of G. Then $UV = G$.
Let $x \in G$. Then xV^{-1} and U are both open and dense in G, hence they have a non-empty intersection. This means that $x \in UV$.

Notice that if G is connected, any non-empty open subset is dense. In that case we need only require U and V to be open and non-empty.

2.2.4. Proposition. Let H be a subgroup of G.

(i) Its closure $\overline{H}$ is a subgroup of G;

(ii) If H contains a non-empty open subset of $\overline{H}$ then $\overline{H} = H$.

Let $x \in H$. Then $X = xH \subset x\overline{H}$. Since $x\overline{H}$ is closed, we have $\overline{H} \subset x\overline{H}$, $x^{-1}\overline{H} \subset \overline{H}$, and $H.\overline{H} \subset \overline{H}$. Thus for each $x \in \overline{H}$ we have $Hx \subset \overline{H}$ and $\overline{Hx} = \overline{H}x \subset \overline{H}$, whence $\overline{H}.\overline{H} \subset \overline{H}$. Also, $(\overline{H})^{-1} = \overline{H^{-1}} = \overline{H}$, and we conclude that $\overline{H}$ is a group. If $U \subset \overline{H}$ is open and contained in H, then H, being a union of translates of U, is also open in $\overline{H}$. By 2.2.3 we have $\overline{H} = H.H = H$.

2.2.5. Proposition. Let ϕ: $G \to G'$ be a homomorphism of algebraic groups.

(i) Ker ϕ is a closed normal subgroup of G;

(ii) ϕG is a closed subgroup of G';

(iii) $\phi(G^0) = \phi(G)^0$.

Ker $\phi = \phi^{-1}e$ is closed in G, whence (i). By 1.9.5 we know that ϕG contains a non-empty open piece of its closure $\overline{\phi G}$. Now (ii) follows from 2.2.4(ii).

$\phi(G^0)$ is a closed subgroup of G' by (ii), which is connected (by 1.2.3(ii)), and of finite index in ϕG, since G^0 is of finite index in G. Then (iii) follows by using 2.2.1(iii).

2.3. G-spaces.

2.3.1. A G-variety, or a G-space, is a variety X, on which G acts as a permutation group, the action being given by a morphism of varieties.

More precisely, there is a morphism of varieties $\alpha: G \times X \to X$, written $\alpha(g,x) = g.x$, such that $g.(h.x) = (gh).x$, $e.x = x$.

X is a homogeneous space for G if it is a G-space on which G acts transitively.

Let X and Y be G-spaces. A morphism $\phi: X \to Y$ is called a G-morphism, or is said to be equivariant, if $\phi(g.x) = g.\phi x$ $(g \in G, x \in X)$. If, moreover, Y is a homogeneous space, then ϕ is clearly surjective.

Let X be a G-space. If $x \in X$ its orbit is the set $G.x = \{g.x | g \in G\}$. It is clear that G acts transitively on G.x. In fact, G.x is a homogeneous space: this follows from 4.3.1(i) see 4.3.5(2). The isotropy group of x is the closed subgroup $G_x = \{g \in G | g.x = x\}$ (check that G_x is closed).

As an example of a G-space, take X = G, and let G act by inner automorphisms: $\alpha(g,x) = gxg^{-1}$ (this is not a homogeneous space, if $G \neq \{e\}$). An example of a homogeneous space is X = G, with G acting by left (right) translations: $\alpha(g,x) = gx$ (gx, respectively). This is even a principal homogeneous space, where G acts simply transitively.

We shall also encounter the following situation. Let V be a finite dimensional vector space over k. A homomorphism of algebraic groups $\phi: G \to GL(V)$ (see 2.1.4(1)) is called a rational representation of G (in V). In this case we can view as a G-space, by defining $g.v = \phi(g)v$ $(g \in G, v \in V)$.

We can also view $\mathbb{P}(V)$ (1.7.2(2)) as a G-space.

2.3.2. Exercises. (1) Let G be a closed subgroup of $\mathbb{GL}_n$ ($n \geq 1$). Then $\mathbb{A}^n$ can be made into a G-space, which is not homogeneous.

(2) The action of $G = \mathbb{GL}_2$ on $\mathbb{A}^2$ defines an action of G on $\mathbb{P}^1$ which makes $\mathbb{P}^1$ into a homogeneous space. Describe the isotropy group of a point. The induced action of G in $\mathbb{P}^1 \times \mathbb{P}^1$ is not homogeneous, in fact, G has 2 orbits in $\mathbb{P}^1 \times \mathbb{P}^1$.

(3) Generalize the results of (2) to $\mathbb{GL}_n$ and $\mathbb{P}^{n-1}$.

2.3.3. From now on we assume that G is a linear algebraic group. Let X be an affine G-space, assume that $\alpha: G \times X \to X$ defines the action.

We have $k[G \times X] = k[G] \otimes_k k[X]$, and α defines an algebra homomorphism $\alpha^*: k[X] \to k[G] \otimes_k k[X]$.

Define a representation σ of G in the (in general infinite dimensional) vector space k[X] by

$$(\sigma(g)f)(x) = f(g^{-1}x).$$

From the next result we see that the representation σ is "locally finite", i.e. that k[X] is a union of finite dimensional subspace which are $\sigma(G)$-stable.

2.3.4. Proposition. Let F be a finite dimensional subspace of k[X].

(i) There is a finite dimensional subspace E of k[X] which contains F and is stable under all $\sigma(g)$ $(g \in G)$;

(ii) F is stable under all $\sigma(g)$ if and only if $\alpha^*F \subset k[G] \otimes_k F$.

To prove (i) we may assume that F = kf. Let

$$\alpha^*f = \sum_{i=1}^{n} u_i \otimes f_i \quad (u_i \in k[G],\ f_i \in k[X]).$$

Then

$$\sigma(g)f(x) = f(g^{-1}x) = \sum_{i=1}^{n} u_i(g^{-1})f_i(x),$$

and we see that all $\sigma(g)f$ lie in the subspace of $k[X]$ spanned by the f_i. The subspace E of E' spanned by all $\sigma(g)f$ $(g \in G)$ satisfies the requirements of (i). If $\alpha^*F \subset k[G] \otimes F$, a similar argument shows that F is $\sigma(G)$-stable.

Next assume that F is $\sigma(G)$-stable. Let (f_i) be a basis of F, and extend it to a basis $(f_i) \cup (g_j)$ of $k[X]$. If $f \in F$, write

$$\alpha^*f = \sum u_i \otimes f_i + \sum v_j \otimes g_j \quad (u_i, v_j \in k[G]),$$

then

$$\sigma(g)f = \sum u_i(g^{-1})f_i + \sum v_j(g^{-1})g_j,$$

and our assumption implies that $v_j(g^{-1}) = 0$ for all g, hence $v_j = 0$. This establishes (ii).

We now consider the case that $X = G$, and that G acts by left - or right - translations. If $f \in k[G]$ put

$$\lambda(g)f(x) = f(g^{-1}x),\ \rho(g)f(x) = f(xg),$$

where $g,x \in G$.

Then $\lambda(gh) = \lambda(g)\lambda(h)$, $\rho(gh) = \rho(g)\rho(h)$ $(g,h \in G)$. So λ and ρ are representations of G in the vector space $k[G]$. These representations are faithful: if $\lambda(g) = \mathrm{id}$ then $f(g^{-1}) = f(e)$ for all $f \in k[G]$, whence $g = e$. Similarly for ρ.

If $\iota x = x^{-1}$, then $\rho(g) = \iota^*\lambda(g)(\iota^*)^{-1}$.

2.3.5. Theorem. Let G be a linear algebraic group. There is an isomorphism of G onto a closed subgroup of some $\mathbb{GL}_n$.

By 2.3.4 we may assume that $k[G] = k[f_1,\dots,f_n]$, where $E = kf_1 + \dots + kf_n$ is $\rho(G)$-stable. We may also assume that (f_i) is a basis of E. By 2.3.4(ii) we have

$$(1) \quad \rho(x)f_i = \sum_{j=1}^{n} m_{ji}(x)f_j,$$

where $m_{ij} \in k[G]$. Then $\phi(x) = (m_{ij}(x))_{1\leq i,j\leq n}$ defines a group homomorphism $\phi\colon G \to \mathbb{GL}_n$.

ϕ is injective: if $\rho(x)f_i = f_i$ $(1 \leq i \leq n)$ then $f_i(x) = f_i(e)$ for all i, whence $f(x) = f(e)$ for all $f \in k[G]$, and $x = e$.

Also, ϕ is a morphism of affine varieties: the corresponding algebra homomorphism $\phi^*\colon k[\mathbb{GL}_n] = k[T_{ij},D^{-1}] \to k[G]$ (notations of 2.1.3(3)) is given by $\phi^*T_{ij} = m_{ij}$, $\phi^*(D^{-1}) = (\det(m_{ij}))^{-1}$.

From (1) we also see that

$$f_i(x) = \sum_{j=1}^{n} m_{ji}(x)f_j(e),$$

from which we conclude that ϕ^* is surjective.

By 2.2.5(ii), G is a closed subgroup of $\mathbb{GL}_n$, and ϕ defines a group isomorphism $G \tilde{\to} \phi G$. Moreover, ϕ^* defines (by 1.10.1 (ii)) a bijective map of $k[\mathbb{GL}_n]/I_{\mathbb{GL}_n}(\phi G)$ onto $k[G]$. It follows that ϕ defines an isomorphism of algebraic groups $G \tilde{\to} \phi G$.

2.3.6. Lemma. Let G be a linear algebraic group, let H be a closed subgroup. Then

$$H = \{x \in G \mid \lambda(x)I_G(H) = I_G(H)\} =$$

$$\{x \in G \mid \rho(x)I_G(H) = I_G(H)\}.$$

It suffices to prove the assertion for λ. If $x,y \in H$, $f \in I_G(H)$, then $\lambda(x)f(y) = f(x^{-1}y)$, so $\lambda(x)f \in I_G(H)$. Conversely, if $\lambda(x)I_G(H) \subset I_G(H)$, then $f(x^{-1}) = \lambda(x)f(e) = 0$ for all $f \in I_G(H)$. Hence $x^{-1} \in H$, and $x \in H$.

2.3.7. Exercise. Let G be a linear algebraic group and X an affine G-variety. Show that there is an isomorphism ψ of X onto a closed subvariety of some $\mathbb{A}^n$, and a rational representation $\phi: G \to \mathbb{GL}_n$ such that $\psi(g.x) = \phi(g)\psi(x)$ $(g \in G, x \in X)$ (Hint: adapt the proof of 2.3.5).

2.4. Jordan decomposition.

2.4.1. The Jordan decomposition is an indispensible tool in the study of linear algebraic groups. We begin by recalling the results from linear algebra which are the basis of it. Let V be a finite dimensional vector space over k. An endomorphism $a \in V$ is semi-simple if there is a basis of V consisting of eigenvectors of a. We say that a is nilpotent if $a^N = 0$ for some N and that a is unipotent if a-1 is nilpotent. Notice that if the characteristic p of k is nonzero, a is unipotent if and only if $a^{p^s} = 1$ for some $s > 0$. We denote by End(V) the algebra of endomorphisms of V; GL(V) is the group of invertible endomorphisms of V. If dim V = n, we may view End(V) as the algebra $\mathbb{M}_n$ of all $n\times n$-matrices and GL(V) as $\mathbb{GL}_n$.

The next lemma is a well-known result from linear algebra.

2.4.2. _Lemma_. _Let_ $S \subset \mathbb{M}_n$ _be a set of commuting matrices._
(i) _There is_ $x \in \mathbb{GL}_n$ _such that_ xSx^{-1} _consists of upper triangular matrices;_
(ii) _If all elements of_ S _are semi-simple, there is_ $x \in \mathbb{GL}_n$ _such that_ xSx^{-1} _consists of diagonal matrices._
We may assume that S is a linear subspace of $\mathbb{M}_n$. Fix a nonzero $s \in S$, and let $W \subset k^n$ be a nonzero eigenspace for s. If all s are scalar multiplications the assertions are trivial. Otherwise there is $s \in S$ such that W is a proper subspace, which is S-stable, because of the assumption about S. Induction on n then implies (i).
(ii) is proved similarly, writing V as a direct sum of eigenspaces for s.

2.4.3. _Lemma_. (i) _The product of two commuting semi-simple_ (_nilpotent_, _unipotent_) _endomorphisms of_ V _is semi-simple_ (_nilpotent_, _unipotent_, _respectively_);
(ii) _If_ $a \in End(V)$, $b \in End(W)$ _are semi-simple_ (_nilpotent_, _unipotent_) _then the same is true for_ $a \oplus b \in End(V \oplus W)$ _and_ $a \otimes b \in End(V \otimes W)$, $a \otimes 1 + 1 \otimes b \in End(V \otimes W)$.
The assertion about semi-simple automorphisms in (i) follows from 2.4.2(ii). The easy proofs of the other statements are left to the reader.

2.4.4. _Proposition_. _Let_ $a \in End(V)$.
(i) _There are unique_ $a_s, a_n \in End(V)$ _such that_ a_s _is semi-simple_, a_n _is nilpotent_, $a_s a_n = a_n a_s$ _and_ $a = a_s + a_n$ (_additive_

<u>Jordan decomposition</u>);

(ii) <u>There are polynomials</u> $P,Q \in k[T]$ <u>without constant term such that</u> $a_s = P(a)$, $a_n = Q(a)$;

(iii) <u>If</u> $b \in End(V)$ <u>commutes with</u> a, <u>then it also commutes with</u> a_s <u>and</u> a_n;

(iv) <u>If</u> $W \subset V$ <u>is an</u> a-<u>stable subspace of</u> V, <u>then</u> W <u>is also stable under</u> a_s <u>and</u> a_n, <u>and</u> $a|W = a_s|W + a_n|W$ <u>is the additive Jordan decomposition of the induced endomorphism</u> $a|W$. <u>A similar results hold for</u> V/W.

Let $\det(T.1-a) = \Pi(T-a_i)^{n_i}$ be the characteristic polynomial of a, the a_i being the distinct eigenvalues of a. Put

$$V_i = \{x \in V | (a-a_i)^{n_i} x = 0\}.$$

The V_i are a-stable subspaces, and V is their direct sum. By the Chinese remainder theorem there exists $P \in k[T]$ with

$$P(T) \equiv a_i \bmod (T-a_i)^{n_i},\ P(T) \equiv 0 \pmod T.$$

Put $a_s = P(a)$. Then a_s stabilizes all V_i, and $a_s|V_i = a_i.\mathrm{id}$. It follows that a_s is semi-simple and $a-a_s$ is nilpotent. The statements of 2.4.4 now readily follow, except for the uniqueness in (i). To prove it, let $a = b_s + b_n$ be a second decomposition (b_s semi-simple, b_n nilpotent, commuting with a). From (iii) we see that b_s and b_n commute with a_s and a_n. From 2.4.3(i) we conclude that $a_s-b_s = b_n-a_n$ is both semi-simple and nilpotent, hence must be 0.

2.4.5. <u>Corollary</u>. <u>Let</u> $a \in GL(V)$. <u>There are unique</u> $a_s, a_u \in GL(V)$ <u>such that</u> a_s <u>is semi-simple</u>, a_u <u>is unipotent</u>, $a_s a_u = a_u a_s$ <u>and</u>

$a = a_s a_u$ (*multiplicative Jordan decomposition*). *We have properties similar to those of* 2.4.3(iii), (iv), *with* a_u *replaced by* a_u.

Let $a = a_s + a_n$ be the additive Jordan decomposition of a. Since a is invertible, it has no eigenvalue 0, and from the proof of 2.4 we see that a_s is invertible. Now $a_u = 1 + a_s^{-1} a_n$ is as required.

We always write $a = a_s + a_n$, $a = a_s a_u$ for the additive and multiplicative Jordan decompositions of $a \in End(V)$, resp. $a \in GL(V)$. We call a_s, a_n, a_u the semi-simple, nilpotent, unipotent parts of a, respectively.

2.4.6. *Corollary*. *Let* $a = a_s a_u \in GL(V)$, $b = b_s b_u \in GL(W)$. *Then* $a \oplus b = (a_s \oplus b_s)(a_u \oplus b_u)$ *is the Jordan decomposition of* $a \oplus b \in GL(V \oplus W)$, *and* $a \otimes b = (a_s \otimes b_s)(a_u \otimes b_u)$ *that of* $a \otimes b \in GL(V \otimes W)$.

This follows by using 2.4.3(ii).

2.4.7. Let V be a not necessarily finite dimensional vector space over k. Denote again by End(V) and GL(V) the algebra of endomorphisms of V and the group of invertible endomorphisms, respectively. We say that $a \in End(V)$ is locally finite if V is a union of finite dimensional a-stable subspaces. We say that a is semi-simple (locally nilpotent) if its restriction to any finite dimensional a-stable subspace is semi-simple (nilpotent). If a is semi-simple it is also semi-simple in the sense of 2.4.1. (the analogous statement for "nilpotent" is not true). For a locally finite a we again have an additive Jordan decomposition $a = a_s + a_n$ satisfying 2.4(i), with locally finite a_s and a_n. To define $a_s x$ and $a_n x$ for $x \in V$, take a finite dimensional a-stable subspace W containing x and define

$$a_s x = (a|W)_s x, \quad a_n x = (a|W)_n x.$$

It follows from 2.4.4 that this is independent of the choice of W and that a_s and a_n are as required.
Similarly, we have multiplicative Jordan decomposition $a = a_s a_u$, if $a \in GL(V)$ is locally finite. The properties of 2.4.4(iv) and 2.4.5 continue to hold, and the same is true for 2.4.4(iii), if b is locally finite.

Now let G be a linear algebraic group, and let $k[G] = A$. If $x \in G$ then the right translation $\rho(x)$ is a locally finite element of GL(A) (by 2.3.4), hence we have a Jordan decomposition

$$\rho(x) = \rho(x)_s \rho(x)_u.$$

2.4.8. <u>Theorem</u>. (i) (<u>Jordan decomposition in</u> G) <u>There are unique</u> $x_s, x_u \in G$ <u>such that</u> $\rho(x)_s = \rho(x_s)$, $\rho(x)_u = \rho(x_u)$, <u>and</u> $x = x_s x_u = x_u x_s$;
(ii) <u>If</u> $\phi: G \to G'$ <u>is a homomorphism of algebraic groups then</u> $\phi(x)_s = \phi(x_s)$, $\phi(x)_u = \phi(x_u)$;
(iii) <u>If</u> $G = \mathbf{GL}_n$ <u>then</u> x_s <u>and</u> x_u <u>are the semi-simple and unipotent parts of</u> 2.4.4.
The elements x_s and x_u of (i) are called the semi-simple and unipotent parts of x. To prove (i), we start from the obser-

vation that $\rho(x)$ is an algebra automorphism of A. Let $q: A \otimes A \to A$ be the surjective homomorphism with $q(a \otimes b) = ab$. That $\rho(x) \in GL(A)$ is an automorphism is expressed by

$$q \circ (\rho(x) \otimes \rho(x)) = \rho(x) \circ q.$$

Using 2.4.4(iv) and 2.4.6 it follows (check this) that we have

$$q \circ (\rho(x)_s \otimes \rho(x)_s) = \rho(x)_s \circ q,$$

which shows that $\rho(x)_s$ is also an automorphism of A. Hence the map $f \mapsto (\rho(x)_s f)(e)$ defines a homomorphism $A \to k$, i.e. a point x_s of G, such that $(\rho(x)_s f)(e) = f(x_s)$ for all $f \in A$.
Since $\rho(x)$ commutes with all left translations $\lambda(y)$ (which are locally finite) we have by 2.4.4(iii)

$$(\rho(x)_s f)(y) = (\lambda(y^{-1})\rho(x)_s f)(e) = (\rho(x)_s \lambda(y^{-1})f)(e) =$$

$$\lambda(y^{-1})f(x_s) = f(yx_s),$$

and we see that $\rho(x)_s = \rho(x_s)$.
We define x_u similarly, and we have $\rho(x)_u = \rho(x_u)$. Then $\rho(x_s x_u) = \rho(x_s)\rho(x_u) = \rho(x)_s\rho(x)_u = \rho(x)_u\rho(x)_s = \rho(x_u x_s)$.
Since ρ is faithful it follows that $x_s x_u = x_u x_s$. We have now established the assertions of (i).
A homomorphism of algebraic groups $\phi: G \to G'$ can be factored: $G \to \mathrm{Im}\,\phi \to G'$. Using 2.2.5(ii) it follows that it suffices to prove (ii) in two special cases:
(a) G is a closed subgroup of G', and ϕ is the canonical injection. Then $k[G] = k[G']/I$, and by 2.3.6

$$G = \{x \in G' \mid \rho(x)I = I\}.$$

The assertion of (ii) now follows by using 2.4.4(iv).

(b) ϕ is surjective. In this case, $k[G']$ can be viewed as a subspace of $k[G]$ (see 1.9.1(i)), which is invariant under all $\rho(x)$, $x \in G$. Again, the assertion follows from 2.4.4(iv).

If $G = \mathbf{GL}_n$, then $A = k[T_{ij}, D^{-1}]$ (see 2.1.3(3)). Put $e_i = T_{1i}$. If $x = (x_{ij}) \in G$, then

$$\rho(x)e_i = \sum_{j=1}^{n} x_{ji}e_j.$$

Hence the subspace $V = ke_1 + \ldots + ke_n$ of A is $\rho(G)$-stable and the restriction of $\rho(x)$ to V is represented by the matrix x, on the basis (e_i). It follows that, x_s and x_u denoting the semi-simple and unipotent parts in $\mathbf{GL}_n$,

$$\rho(x)_s|V = \rho(x_s)|V, \ \rho(x)_u|V = \rho(x_u)|V.$$

This implies (iii).

2.4.9. Corollary. *$x \in G$ is semi-simple (unipotent) if and only if for any isomorphism ϕ of G onto a closed subgroup of some $\mathbf{GL}_n$ we have that $\phi(x)$ is semi-simple (resp. unipotent).*

2.4.10. Exercises. Notations of 2.4.8.

(1) Show that $\lambda(x)_s = \lambda(x_s)$, $\lambda(x)_u = \lambda(x_u)$.

(2) The set G_u of unipotent elements of G is a closed subset.

(3) Show by an example that the set G_s of semi-simple elements of G is not necessarily open or closed.

The linear algebraic group G is *unipotent* if all its elements are unipotent (warning: a semi-simple linear algebraic group is <u>not</u> one all whose elements are semi-simple, see the definition in 6.14).

The next result, which is in fact one from linear algebra, together with 2.3.5 and 2.4.9, shows that a unipotent G is isomorphic to a closed subgroup of some group $\mathbb{U}_n$ of unipotent upper triangular matrices.

2.4.11. *Theorem. Let G be a subgroup of* $\mathbb{GL}_n$ *consisting of unipotent matrices. There is* $x \in \mathbb{GL}_n$ *such that* $xGx^{-1} \subset \mathbb{U}_n$.

Let $V = k^n$, so G is a subgroup of $GL(V) = \mathbb{GL}_n$. Use induction on n. It suffices to prove that there is $v \in V$, $v \neq 0$, such that $gv = v$ for all $g \in G$. If G stabilizes a nontrivial proper subspace of V the existance of v follows by induction. So we are reduced to the case that G acts irreducibly in V. Let $A \subset \mathbb{M}_n$ be the subspace spanned by all $g \in G$: this is a subalgebra of the algebra $\mathbb{M}_n$, which acts irreducibly in V. By Burnside's theorem (see e.g. [26 , Ch. XVII, §3]) it follows that $A = \mathbb{M}_n$. If $g,h \in G$ then $Tr((1-g)h) = Tr\,h - Tr(gh) = 0$. Hence $Tr((1-g)a) = 0$ for all $a \in A = \mathbb{M}_n$. But then we must have $g = 1$, $G = \{1\}$, $\dim V = 1$, and the existence of v is trivial.

In the rest of this chapter we shall deal with various kinds of abelian groups. First we prove a structure theorem for abelian linear algebraic groups.

2.4.12. *Theorem. Let G be a commutative linear algebraic group.*

(i) _The sets G_s, G_u of semi-simple and unipotent elements are closed subgroups_;

(ii) _The product map $\pi: G_s \times G_u \to G$ is an isomorphism of algebraic groups_.

We may assume G to be a closed subgroup of $\mathbb{GL}_n$. From 2.4.3(i) we see that G_s and G_u are subgroups, moreover G_u is closed (2.4.10(2)). It follows from 2.4.2(ii) that there exists a direct sum decomposition of $V = k^n$: $V = \oplus V_i$, and homomorphisms $\phi_i: G_s \to k^*$ such that $g.v = \phi_i(g)v$ for $g \in G_s$, $v \in V_i$. The V_i are clearly G-stable. Trigonalizing the restrictions of G to V_i (2.4.2(i)) we may assume that $G \subset \mathbb{T}_n$, $G_s = G \cap \mathbb{D}_n$ (notations of 2.1.3). Then G_s is closed, whence (i).

The uniqueness of the Jordan decomposition in G implies that π is an isomorphism of abstract groups. It is also clear that π is a morphism of algebraic varieties. To establish (ii), it suffices to show that π^{-1} is also a morphism. But the map $x \mapsto x_s$ is a morphism, being "projection onto diagonal matrices". Hence $\pi: x \mapsto (x_s, xx_s^{-1})$ is also a morphism.

2.4.13. _Corollary_. _If, moreover_, G _is connected then the same is true for G_s and G_u_.

2.5. _Diagonalizable groups and tori_.

2.5.1. Let G be a linear algebraic group. A homomorphism of algebraic groups $\chi: G \to \mathbb{G}_m$ is called a _rational character_ of G. The set of rational characters is denoted by $X^*(G)$, it has a natural structure of abelian group. Its group operation is often written additively. We may view $X^*(G)$ as a subset of k[G].

We denote by $X_*(G)$ the set of homomorphisms of algebraic groups $\lambda: \mathbb{G}_m \to G$. Such a λ is called a multiplicative 1-parameter subgroup of G. If G is commutative then $X_*(G)$ has also a structure of additive abelian group: we define $(\lambda+\mu)(t) = \lambda(t)\mu(t), (-\lambda)(t) = \lambda(t)^{-1}$ $(t \in \mathbb{G}_m)$.

A linear algebraic group G which is isomorphic to a closed subgroup of some group of diagonal matrices $\mathbb{D}_n$ is called diagonalizable. G is an algebraic torus (or simply a torus) if it is isomorphic to some $\mathbb{D}_n$.

2.5.2. Theorem. *The following properties of a linear algebraic group G are equivalent:*

(a) G is diagonalizable;

(b) $X^*(G)$ is an abelian group of finite type, its elements generate k[G];

(c) Any rational representation of G is a direct sum of 1-dimensional ones.

Write an element $d \in \mathbb{D}_n$ as $\operatorname{diag}(\chi_1(d),\ldots,\chi_n(d))$. Then χ_i is a rational character of $\mathbb{D}_n$, and we have $k[\mathbb{D}_n] = k[\chi_1,\ldots,\chi_n,\chi_1^{-1},\ldots,\chi_n^{-1}]$. We see that $X^*(\mathbb{D}_n)$ generates $k[\mathbb{D}_n]$. If G is a closed subgroup of $\mathbb{D}_n$, the k[G] is a quotient of $k[\mathbb{D}_n]$, and it is clear that $X^*(G)$ generates k[G].
In fact, ϕ_i denoting the image of χ_i in k[G], the rational characters $\phi_1^{a_1} \ldots \phi_n^{a_n}$ of G span k[G], hence any rational character is a linear combination of them. But it then follows from Dedekind's theorem on linear dependence of characters (see [26 , Ch. VIII,§4]) that the characters $\phi_1^{a_1} \ldots \phi_n^{a_n}$

exhaust $X^*(G)$. Hence $X^*(G)$ is an abelian group of finite type, proving (b).

If (b) holds, the $\chi \in X^*(G)$ generate $k[G]$, and by Dedekind's theorem they form a basis. Let $\phi: G \to \mathbb{GL}_n$ be a rational representation. Put $V = k^n$. Expressing the matrix elements of $\psi(g)$ as linear combinations of elements of $X^*(G)$, we see that we can write $\phi(g)$ as a finite sum

$$\phi(g) = \sum_{\chi \in X^*(G)} \chi(g) A_\chi,$$

where $A_\chi \in \mathbb{M}_n$. From $\phi(gh) = \phi(g)\phi(h)$ we infer

$$\sum_\chi \chi(g)\chi(h)A_\chi = \sum_{\chi,\chi'} \chi(g)\chi'(h)A_\chi A_{\chi'}.$$

Application of Dedekind's theorem to $G \times G$ shows that

$$A_\chi A_{\chi'} = \delta_{\chi,\chi'} A_\chi,$$

and we also have

$$\sum_\chi A_\chi = 1.$$

Put $V_\chi = A_\chi V$. Then V is the direct sum of the nonzero V_χ, and

$$\phi(g)v = \chi(g)v,$$

if $g \in G$, $v \in V_\chi$. This proves (c).

That (c) implies (a) follows by imbedding G in some $\mathbb{GL}_n$, and using (c).

2.5.3. Corollary. *Let H be a closed subgroup of the diagonalizable group G. Then H is diagonalizable, and it is the*

<u>intersection of the kernels of finitely many rational characters of</u> G.

The first point is clear. Let $f \in k[G]$ vanish on H. Writing f as a linear combination of the $\chi \in X^*(G)$, and using Dedekind's theorem, we see that f lies in the ideal generated by the $\chi - 1$, where $\chi(H) = 1$. Since these χ form a subgroup of finite type, the last assertion follows.

2.5.4. <u>Corollary</u>. <u>Let</u> G <u>be diagonalizable</u>. $X^*(G)$ <u>is an abelian group of finite type</u>, <u>without</u> p-<u>torsion</u>, <u>if</u> $p = \text{char } k > 0$. <u>The algebra</u> $k[G]$ <u>is isomorphic to the group algebra of</u> $X^*(G)$.

The first point follows from (b), using that k contains no p^{th} roots of unity $\neq 1$, if $p > 0$. The second point has been established in the course of the proof of 2.5.3 (see also 2.5.6)

2.5.5. <u>Exercises</u>. (1) Determine the rational characters of $\mathbb{G}_m$.

(2) Let G be a linear group. G acts on $X^*(G)$ by $(g.\lambda)(t) = g\lambda(t)g^{-1}$ $(t \in \mathbb{G}_m)$.

If $g \in G$, $\lambda \in X_*(G)$, define $\theta_{g,\lambda}: \mathbb{G}_m \to G$ by $\theta_{g,\lambda}(t) = \lambda(t)g\lambda(t)^{-1}$. This is a morphism of algebraic varieties. Viewing $\mathbb{G}_m$ as an open subset of the affine line, let $P(\lambda)$ be the subset of the $g \in G$ such that $\theta_{g,\lambda}$ extends to a morphism $\mathbb{A}^1 \to G$.

(a) $P(\lambda)$ is a subgroup of G, containing the centralizer H in G of $\lambda(\mathbb{G}_m)$. We have $P(g.\lambda) = gP(\lambda)g^{-1}$;

(b) $P(\lambda) \cap P(-\lambda) = H$ (Hint: use 1.6.12(2), last part);

(c) Let $G = \mathbb{GL}_n$, acting in $V = k^n$. For any $\lambda \in X_*(G)$ there exists a family of subspaces $V_1 \subset V_2 \subset \dots \subset V_s \subset V$ (a "flag") such that

$$P(\lambda) = \{g \in G | g.V_i = V_i, 1 \leqslant i \leqslant s\}.$$

(Hint: using 2.5.2(c), reduce to the case that $\lambda(t) = \mathrm{diag}(t^{a_1},\dots,t^{a_n})$, where $a_1 \geqslant a_2 \geqslant \dots \geqslant a_n$);
(d) Show that $P(\lambda)$ is a closed subgroup of G, for any G and $\lambda \in X_*(G)$.

2.5.6. Recall that the group algebra $k[\Gamma]$ of the abelian group Γ (whose group operation is written as addition) is the k-algebra with a basis $(e(\gamma))_{\gamma\in\Gamma}$, the multiplication being defined by

$$e(\gamma)e(\delta) = e(\gamma+\delta).$$

We have

$$k(\Gamma_1 \oplus \Gamma_2) \simeq k[\Gamma_1] \otimes_k k[\Gamma_n].$$

Recall that a Γ of finite type is a direct sum of (finite or infinite) cyclic groups (see [26 , Ch. I, §10]).
Now let Γ be an abelian group of finite type, without p-torsion (if p = char k > 0). Define $\pi^*: k[\Gamma] \to k[\Gamma] \otimes_k k[\Gamma]$, $\iota^*: k[\Gamma] \to k[\Gamma]$ by $\pi^* e(\gamma) = e(\gamma) \otimes e(\gamma)$, $\iota^* e(\gamma) = e(-\gamma)$, and let e be the homomorphism $k[\Gamma] \to k$ sending all $e(\gamma)$ to 1.

2.5.7. <u>Proposition</u>. (i) $k[\Gamma]$ <u>is an affine</u> k-<u>algebra</u>;
(ii) <u>There is a diagonalizable group</u> $G(\Gamma)$ <u>with</u> $k[G(\Gamma)] = k[\Gamma]$, <u>such that</u> π^*, ι^*, e <u>determine multiplication, inversion and identity element of</u> $G(\Gamma)$;

(iii) $X^*(\mathcal{G}(\Gamma)) \simeq \Gamma$; *if* G *is diagonalizable then* $\mathcal{G}(X^*(G)) \simeq G$.
To prove (i) it suffices to show that $k[\Gamma]$ is reduced. Writing Γ as a direct sum of cyclic groups and using 1.5.2 we see that it suffices to consider the case of a cyclic Γ. If Γ is infinite cyclic then $k[\Gamma] \simeq k[T,T^{-1}]$, an integral domain. If Γ is finite, then $k[\Gamma] \simeq k[T]/(T^d-1)$, where p does not divide d (if $p > 0$). This is a reduced algebra, since the polynomial T^d-1 has no multiple roots. The proof of (ii) and (iii) can also be given by reducing it to the case of a cyclic Γ. We leave the details to the reader (use that $\mathcal{G}(\mathbb{Z}) \simeq \mathbb{G}_m$, $\mathcal{G}(\mathbb{Z}/m\mathbb{Z}) \simeq \mathbb{Z}/m\mathbb{Z}$, $X^*(\mathbb{G}_m) \simeq \mathbb{Z}$, $X^*(\mathbb{Z}/m\mathbb{Z}) \simeq \mathbb{Z}/m\mathbb{Z}$).
The isomorphisms of (iii) are functorial, in the appropriate categories (cf. 2.5.9(1)).

2.5.8. *Proposition*. *Let* G *be a diagonalizable group*.
(i) G *is a direct product of a torus and a finite abelian group*;
(ii) G *is a torus if and only if* G *is connected*;
(iii) G *is a torus if and only* $X^*(G)$ *is a free abelian group*.
Write $X^*(G)$ as a direct sum of a free abelian group Γ_1 and a finite one Γ_2. From 2.5.7(iii) we find that $G \simeq \mathcal{G}(\Gamma_1) \times \mathcal{G}(\Gamma_2)$. Since $k[\Gamma_2]$ is finite dimensional, $\mathcal{G}(\Gamma_2)$ must be finite (e.g. by using 1.8.4(2)). Hence if G is connected then $\mathcal{G}(\Gamma_2) = 1$, i.e. $\Gamma_2 = 0$.
The assertions now all follow by observing that $X^*(\mathbb{D})_n \simeq \mathbb{Z}^n$ and $\mathcal{G}(\mathbb{Z}^n) = \mathcal{G}(\mathbb{Z}) \times \ldots \times \mathcal{G}(\mathbb{Z}) \simeq \mathbb{D}_n$, using again 2.5.7(iii).

2.5.9. *Exercises*. (1) Make diagonalizable groups and abelian groups without p-torsion into categories and show that these two categories are anti-equivalent.

(2) Let G be diagonalizable. If H is a closed subgroup of G and Y a subgroup of X*(G), define

$$H^0 = \{\chi \in X^*(G) \mid \chi(H) = 1\},$$

$$Y^0 = \{x \in G \mid \chi(x) = 1, \text{ for all } y \in Y\}.$$

Then $H^{00} = H$ and $Y^{00} = Y$ if $X^*(G)/Y$ has no p-torsion.

We conclude this section with some special results about diagonalizable groups and tori (partly in the form of exercises).

2.5.10. Proposition. ("rigidity of diagonalizable groups"). *Let G and G' be diagonalizable groups, let V be a connected affine variety. Let $\phi: V \times G \to G$ be a morphism of affine varieties such that for any $v \in V$, the map $x \mapsto \phi(v,x)$ defines a homomorphism of algebraic groups. Then $\phi(v,x)$ is independent of v.*

If $\chi' \in X^*(G')$, we have that $\chi'\phi(v,g)$ can be written in the form

$$\chi'\phi(v,g) = \sum_{\chi \in X^*(G)} f_{\chi,\chi'}(v)\chi(g),$$

with $f_{\chi,\chi'} \in k[V]$. An argument similar to one used in the proof of 2.5.3 shows that

$$f_{\chi,\chi'} f_{\chi_1,\chi'} = \delta_{\chi,\chi_1} f_{\chi,\chi'} \quad (\chi,\chi_1 \in X^*(G), \chi' \in X^*(G')).$$

In particular, any $f_{\chi,\chi'}$ is an indempotent element of k[V], so takes only the values 0 and 1. For $\varepsilon = 0,1$ put

$V_\varepsilon = \{v \in V | f_{\chi,\chi'}(v) = \varepsilon\}$. Then V_0 and V_1 are open and closed in V, and V is their disjoint union. The connectedness of V implies that $f_{\chi,\chi'}$ is either 0 or 1.
This establishes the assertion.

Let G be a linear algebraic group, and H a closed subgroup. Denote by $Z_G(H)$ and $N_G(H)$ the centralizer and normalizer of H in G, i.e.

$$Z_G(H) = \{g \in G | ghg^{-1} = h \text{ for all } h \in H\},$$

$$Z_G(H) = \{g \in G | gHg^{-1} = H\}.$$

These are closed subgroups of G. For $N_G(H)$ is the set of all $g \in G$ such that $f(ghg^{-1}) = 0$ for all $f \in k[G]$ vanishing on H and all $h \in H$, hence is defined by the vanishing of a family of functions in $k[G]$. That the same is true for $Z_G(H)$ is readily checked.

2.5.11. Corollary. *Let* H *be a diagonalizable subgroup of* G. *Then* $N_G(H^0) = Z_G(H)^0$ *and* $N_G(H)/Z_G(H)$ *is finite*.
The first point follows from 2.5.10, applied to $V = N_G(H)^0$, $G = G' = H$, with $\phi(v,h) = vhv^{-1}$.
The second point then follows by 2.2.1(i).

2.5.12. Exercises (about tori). Let T be a torus. We put $X^* = X^*(T)$, $X_* = X_*(T)$.
(1) Let $T = \mathbb{D}_n$. Then X^* consists of the functions

$$\operatorname{diag}(x_1,\dots,x_n) \mapsto x_1^{a_1}\dots x_n^{a_n},$$

with $(a_1,\dots,a_n) \in \mathbb{Z}^n$ and X_* consists of the maps $\mathbb{G}_m \to \mathbb{D}_n$

with $x \mapsto \mathrm{diag}(x^{b_1},\ldots,x^{b_n})((b_1,\ldots,b_n) \in \mathbb{Z}^n)$.

(2) If $\chi \in X^*$, $\lambda \in X^*$ define $\langle \chi,\lambda \rangle \in \mathbb{Z}$ by

$$\chi(\lambda(x)) = x^{\langle \chi,\lambda \rangle}.$$

Then $\langle\ ,\ \rangle$ defines a perfect pairing between X^* and X_*, i.e. any homomorphism $X^* \to \mathbb{Z}$ is of the form $\chi \mapsto \langle \chi,\lambda \rangle$, $\chi \in X^*$, and similarly for X_*.

(3) Let V be an affine T-space, let $A = k[V]$. Denote by $B = A^T$ the subalgebra of all $f \in A$ such that $f(t.v) = f(v)$ $(t \in T, v \in V)$.

(a) B is an affine subalgebra of A;

(b) If W is an affine variety with $k[W] \simeq B$ then the injection $B \to A$ defines a morphism of affine varieties $\phi: V \to W$ such that $\phi(t.v) = \phi v$ $(t \in T, v \in V)$. Show that the following holds: if W' is an affine variety and ϕ' a morphism $V \to W'$ with $\phi'(t.v) = \phi' v$ $(t \in T, v \in V)$ there is a unique morphism $\psi: W \to W'$ such that $\phi' = \phi \circ \psi$ (Hint: use the decomposition $A = \bigoplus_{\chi \in X^*} A_\chi$, where $A_\chi = \{f \in A \mid f(t.v) = \chi(t)f(v)\}, t \in T, v \in V)$.

(4) <u>An</u> <u>equivariant</u> <u>affine</u> <u>embedding</u> if T is an affine T-space V containing T as an open subvariety such that the action $T \times V \to V$ extends the product map $T \times T \to T$.

(a) There is a finitely generated sub-semigroup S of X^*, which generates X^*, such that $k[V]$ is isomorphic to the semigroup algebra $k[S]$ of S (Hint: view $k[V]$ as a $\rho(T)$-stable subspace of $k[T]$ and use that $k[T]$ is the group algebra $k[X^*]$);

(b) Conversely, for any S with the properties of (a) there is an equivariant affine embedding V such that $k[V] \simeq k[S]$. It is unique up to an isomorphism of T-spaces.

(For details on equivariant embeddings we refer to: G. Kempf et al., Toroidal Embeddings, Lect. Notes in Math. no. 339, Springer Verlag, 1973, Chapter I).

2.6. One dimensional groups.

2.6.1. In this section we are going to prove that a 1-dimensional connected linear algebraic group is isomorphic to $\mathbb{G}_a$ or $\mathbb{G}_m$. This is a nontrivial result.
A geometric argument to prove it, but which requires more knowledge of algebraic geometry then we presuppose here, runs as follows.
Imbed the affine variety G as an open sub-variety in some irreducible smooth projective curve C. So G = C-F, where F is finite and non-empty. The variety G has an infinite group of automorphisms, viz. the left translations by its elements. Extending then to automorphisms of C fixing the points of F one sees that C has an infinite group of automorphisms fixing a point. Then C is isomorphic to the projective line $\mathbb{P}^1$.
(a proof of this using the Jacobian of C is given in [3, p. 250-252], see also [21 , p. 305-306]).
Thus G is an open subset of $\mathbb{P}^1$ and it follows by elementary arguments that G is as described. The important point here is that the automorphism group of the variety $\mathbb{P}^1$ is $\mathbb{PSL}_2$ (see [3 , p. 252-258] for details).
We follow here an elementary approach. The difficult case is that of a unipotent group in characteristic $p > 0$.

We denote in this section by G a 1-dimensional connected linear algebraic group over k. We put p = char k (p may be 0).

2.6.2. Lemma (i) *G is commutative*.
(ii) *If G is not unipotent then G is isomorphic to* $\mathbb{G}_m$;
(iii) *If G is unipotent and* $p > 0$, *then for all* $g \in G$ *we have* $g^p = e$
Fix $g \in G$ and consider the morphism $\phi: x \to xgx^{-1}$ of G into itself. By 1.2.3 the closure $\overline{\phi G}$ is a closed irreducible subset. Using 1.8.2 it follows that either $\phi G = \{e\}$ or $\overline{\phi G} = G$. In the latter case, by 1.9.5, we have that $G - \phi G$ is a finite set. We may view G as a closed subgroup of some $\mathbb{GL}_n$ (2.3.5). If $G - \phi G$ is finite then there are only finitely many possibilities for the characteristic polynomial $\det(T.1-x)$ of $x \in G$. The connectedness of G then implies that the characteristic polynomial must be constant, hence equal to $(T-1)^n$. Thus G is unipotent, and by 2.4.11 and 2.1.4(4) it is solvable. On the other hand, since $g^{-1}\phi G$ lies in the commutator subgroup G' of G, the finiteness of $G - \phi G$ implies that G-G' is finite. Then $\overline{G'} = G$, and 2.2.4(ii) shows that G' = G. But this contradicts the solvability of G. Hence $\phi G = e$. Since g is arbitrary, it follows that G is commutative, proving (i).
By 2.4.12 and 2.4.13, we now have that G is the direct product of the two connected subgroups G_s and G_u. Since dim G = 1, one of them must be trivial (use 1.8.3). If $G = G_s$ it is diagonalizable (2.4.2(ii)) and by 2.5.8(ii) it is a torus, which then must be isomorphic to $\mathbb{G}_m$. This establishes (ii).
In the situation of (iii), the map $g \mapsto g^p$ is a homomorphism of G into itself. By 2.2.5(ii), using that dim G = 1, the image is either {e} or G. However from 2.4.11 (see also

2.4.1) it follows that all elements of G have p-power order, and the second alternative is impossible. This proves (iii).

To deal with the remaining case: G unipotent we need a number of auxiliary results.
A polynomial $P \in k[T_1,\dots,T_m]$ is a p-_polynomial_ if it is a linear combination of terms $T_i^{p^h}$ $(1 \leq i \leq n,\ h \geq 0)$ if $p > 0$ and if it is homogeneous linear for $p = 0$.

2.6.3. _Lemma_. $P \in k[T_1,\dots,T_m]$ _is a_ p-_polynomial if and only if_

(1) $P(T_1+U_1,\dots,T_m+U_m) = P(T_1,\dots,T_m)+P(U_1,\dots,U_m)$, _the_ T_i _and_ U_j _being indeterminates_.
It suffices to prove that a P satisfying (1) is a p-polynomial. Let D_i be partial derivation in $k[T_1,\dots,T_n]$ with respect to T_i. Then (1) implies that D_iP is a constant a_i $(1 \leq i \leq n)$. Put $Q(T_1,\dots,T_n) = P(T_1,\dots,T_n) - \sum_{i=1}^{n} a_iT_i$. Then $D_iQ = 0$ $(1 \leq i \leq n)$. It follows that $Q = 0$ if $p = 0$, whence the assertion in that case. If $p > 0$ it follows that there is $Q_1 \in k[T_1,\dots,T_n]$ such that $Q(T_1,\dots,T_n) = Q_1(T_1^p,\dots,T_n^p)$. Then Q_1 has the property (1). Since $\deg Q_1 < \deg Q$, induction on the degree establishes the lemma.

Let H be a connected linear algebraic group. An _additive function_ α on H is an element of $k[H]$ such that $\alpha(hh') = \alpha(h) + \alpha(h')$ $(h,h' \in H)$. Notice that if $\alpha_1,\dots,\alpha_m$ are additive functions and $P \in k[T_1,\dots,T_m]$ a p-polynomial, then $P(\alpha_1,\dots,\alpha_m)$ is also an additive function.
The additive functions $\alpha_1,\dots,\alpha_m$ are called _algebraically_

<u>dependent</u> (p-<u>dependent</u>) if there is a nonzero polynomial (resp. a p-polynomial) $P \in k[T_1,\dots,T_m]$ such that

$$P(\alpha_1(h),\dots,\alpha_m(h)) = 0 \quad (h \in H).$$

The corresponding notions of independence are clear. If $p = 0$ then p-dependence is the same as linear dependence.

2.6.4. <u>Lemma</u>. (i) <u>If the additive functions</u> $\alpha_1,\dots,\alpha_m$ <u>are algebraically dependent they are</u> p-<u>dependent, so the notions of algebraic dependence and</u> p-<u>dependence of additive functions are the same;</u>
(ii) <u>If the additive functions</u> $\alpha_1,\dots,\alpha_m$ <u>are</u> p-<u>dependent there exist additive functions</u> $\beta_1,\dots,\beta_n$ <u>and</u> p-<u>polynomials</u> $P_i \in k[T_1,\dots,T_n]$ <u>such that</u>: $n < m$, <u>the</u> β_i <u>are</u> p-<u>independent</u>, <u>and</u> $\alpha_i = P_i(\beta_1,\dots,\beta_n)$ $(1 \leq i \leq m)$.
Let $\alpha_1,\dots,\alpha_m$ be algebraically dependent and take a nonzero $P \in k[T_1,\dots,T_m]$ of minimal degree such that

$$P(\alpha_1(h),\dots,\alpha_m(h)) = 0 \quad (h \in H).$$

Then $P(T_1+\alpha_1(h),\dots,T_m+\alpha_m(h)) - P(T_1,\dots,T_m)$ is a polynomial of degree smaller than that of P, giving an algebraic dependence between $\alpha_1,\dots,\alpha_m$. Hence it is 0. Let $Q(T_1,\dots,T_m)$ be a coefficient of some monomial $U_1^{a_1},\dots,U_n^{a_n}$ in $P(T_1+U_1,\dots,T_m+U_m) - P(T_1,\dots,T_m) - P(U_1,\dots,U_m)$. Then $\deg Q < \deg P$ and $Q(\alpha_1(h),\dots,\alpha_m(h)) = 0$ $(h \in H)$. It follows that $Q = 0$. This establishes (i).
In the proof of (ii) first assume $m = 1$. Then $P(\alpha_1) = 0$, $P \in k[T]$ a p-polynomial. It follows that $\alpha_1(H)$ is finite, and

the connectedness of H implies that $\alpha_1 = 0$.
Let $m > 1$, and assume that $\alpha_1,\dots,\alpha_m$ are p-dependent. If $p=0$ they are linearly dependent, and (ii) is well-known. Now let $p > 0$. We may assume, replacing α_1 by a suitable linear combination of $\alpha_1,\dots,\alpha_m$, that some dependence relation has the form

$$(2)\quad \alpha_1+c_1\alpha_1^p+\dots+c_s\alpha_1^{p^s}+l_1(\alpha_2,\dots,\alpha_m)^p+\dots+l_t(\alpha_2,\dots,\alpha_m)^{p^t}=0,$$

where the $l_i \in k[T_2,\dots,T_m]$ are homogeneous linear, and not all 0. Put $\tilde{\alpha}_1 = -c_1^{\frac{1}{p}}\alpha_1 - \dots - c_s^{\frac{1}{p}}\alpha_1^{p^{s-1}} - l_1(\alpha_2,\dots,\alpha_m) - \dots - l_t(\alpha_2,\dots,\alpha_m)^{p^{t-1}}$. This is an additive function. From (2) we find a p-dependence relation for $\tilde{\alpha}_1,\alpha_2,\dots,\alpha_m$, of lower degree then (2). Induction on this degree then establishes (ii).

Next we establish a result about "polynomial cocycles" in $k[T_1,T_2]$. Recall that $p = \operatorname{char} k$. If $p > 0$, put

$$c(T_1,T_2) = \sum_{i=1}^{p-1} \frac{(p-1)\dots(p-i+1)}{i!}\, T_1^{p-1}T_2^i .$$

(this is $p^{-1}\{(T_1+T_2)^p - T_1^p-T_2^p\} \in \mathbb{Z}[T_1,T_2]$, reduced mod p).

2.6.5. <u>Lemma</u>. <u>Let</u> $f \in k[T_1,T_2]$ <u>satisfy</u>

$$(3)\quad f(T_1+T_2,T_3) + f(T_1,T_2) = f(T_1,T_2+T_3) + f(T_2,T_3),$$

$$(4)\qquad f(T_1,T_2) = f(T_2,T_1).$$

<u>Then the following statements hold</u>.

(i) *If* $p = 0$ *there is* $g \in k[T]$ *such that*

$$f(T_1,T_2) = g(T_1+T_2) - g(T_1) - g(T_2);$$

(ii) *If* $p > 0$ *there are* $g \in k[T]$, $a \in k$, $h \geqslant 0$ *such that*

$$f(T_1,T_2) = g(T_1+T_2) - g(T_1) - g(T_2) + ac(T_1,T_2)^{p^h}.$$

If, moreover,

$$(5) \quad \sum_{i=0}^{p-1} f(T,iT) = 0,$$

then $a = 0$.

We may assume f to be homogeneous of degree d. Let D_i be partial derivation with respect to T_i $(i = 1,2)$. From (3) we get

$$(6) \quad (D_1f)(T_1+T_2,T_3) + (D_1f)(T_1,T_2) = (D_1f)(T_1,T_2+T_3),$$

$$(7) \quad (D_1D_2f)(T_1+T_2,T_3) = (D_1D_2f)(T_1,T_2+T_3).$$

From (7) we conclude that D_1D_2f is a multiple of $(T_1+T_2)^{d-2}$.

Let $p > 0$.

If $d \not\equiv 1 \pmod p$ this implies that

$$(D_1f)(T_1,T_2) = b((T_1+T_2)^{d-1} - T_1^{d-1}) + h(T_1,T_2^p),$$

for some $b \in k$ and $h \in k[T_1,T_2]$. Inserting this into (6) we obtain

$$h(T_1+T_2,T_3^p) + h(T_1,T_2^p) = h(T_1,T_2^p+T_3^p).$$

Putting $T_1 = 0$ we find

$$(8) \quad h(T_1,T_2^p) = h(0,T_1^p+T_2^p) - h(0,T_1^p).$$

Hence the degree of $h(T_1,T_2^p)$ is divisible by p. Since $d \not\equiv 1 \pmod p$ we conclude that

$$(10)\quad (D_1f)(T_1,T_2) = b((T_1+T_2)^{d-1} - T_1^{d-1}).$$

If $d \equiv 1 \pmod p$ we must have $D_1D_2f = 0$, whence

$$D_1f(T_1,T_2) = h(T_1,T_2^p)$$

for some $h \in k[T_1,T_2]$. We have again (8), which implies (check this) that (9) also holds in this case.

If $d \not\equiv 0 \pmod p$, we find from (9) that

$$f(T_1,T_2) = a((T_1+T_2)^d - T_1^d - T_2^d) + h_1(T_1^p,T_2),$$

for some $a \in k$, $h_1 \in k[T_1,T_2]$. The symmetry property (4) implies that $h_1 = 0$. This proves that f has the desired form in this case.

If $d \equiv 0 \pmod p$, $d > p$, $b \neq 0$, then D_1f contains a term $T_1^{p-1}T_2{}^{d-p}$ (since in that case the binomial coefficient $\binom{d-1}{p-1}$ is not divisible by p), which is impossible. So $D_1f = 0$ and f must be the p-th power of a polynomial satisfying (3) and (4), and we can use induction on d. There remains the case that $d = p$. Then it follows directly from (9) that f is multiple of c. This finishes the proof of the first part of (ii). The last assertion of (ii) is easy to check.

The proof of (i), which is easier, is left to the reader.

We can now prove the main result of this section.

2.6.6. _Theorem_. _Let_ G _be_ _a_ _connected_ 1-_dimensional_ _linear_ _algebraic_ _group_. _Then_ G _is_ _isomorphic_ _to_ _either_ $\mathbb{G}_a$ _or_ $\mathbb{G}_m$.

By 2.6.2 we may assume that G is unipotent. We have to prove $G \simeq \mathbb{G}_a$, which means (check this) that there is an additive function $\alpha \in k[G]$ such that $k[G] = k[\alpha]$.

We assume that G is a closed subgroup of some group $\mathbb{U}_n$ of unipotent upper triangular matrices (see 2.4.11). Let $f_{ij} \in k[G]$ be the (i,j)-matrix entry function $(1 \leq i < j \leq n)$.

We use induction on n. For n = 1 there is no problem (for $\mathbb{U}_1 \simeq \mathbb{G}_a$). So assume that $n > 1$ and that the assertion has been proved for closed subgroups of $\mathbb{U}_{n-1}$.

There are two homomorphisms of algebraic groups ϕ_1,ϕ_2: $\mathbb{U}_n \to \mathbb{U}_{n-1}$: the first one being obtained by erasing the first row and column of a matrix, the second one by crasing the last row and column.

By induction, there are additive functions $\alpha_1,\alpha_2 \in k[G]$ such that $k[\phi_i G] = k[\alpha_i]$, i = 1,2. This means that the f_{ij} with $i > 1$ lie in $k[\alpha_1]$ and those with $j < n$ in $k[\alpha_2]$. Since dim G = 1, we have that α_1 and α_2 are algebraically dependent. By 2.6.4 there is an additive $\beta \in k[G]$ such that $\alpha_1,\alpha_2 \in k[\beta]$. Hence all f_{ij} with $(i,j) \neq (1,n)$ lie in $k[\beta]$.

From the multiplication rule for matrices we see that there is a unique $f \in k[T_1,T_2]$ such that

$$f_{1n}(xy)-f_{1n}(x)-f_{1n}(y) = f(\beta(x),\beta(y)) \quad (x,y \in G),$$

and it follows that f satisfies (3) and (4). If $p > 0$, (5) also holds, as follows from 2.6.2(iii). Let g be as in 2.6.5. Then

$$\gamma(x) = f_{1n}(x) - g(\beta(x))$$

is again an additive function. By 2.6.4, β and γ are expressible in an additive function α, and we have $k[G] = k[\alpha]$. This proves the theorem.

2.6.7. Exercises.

(1) Let G be a linear algebraic group which is connected, unipotent and commutative. If $p > 0$ assume moreover that $G^p = \{e\}$.

Show that G is isomorphic to a "vector group" $(\mathbb{G}_a)^n$ (Hint: proceed as in the proof of 2.6.6.)

(2) ($p > 0$) Let c be as in 2.6.5. Define a structure of linear algebraic groups on $\mathbb{A}^2$, with product $(x,y).(x',y') = (x+x',y+y'+c(x,x'))$. This group is connected, unipotent, commutative, of dimension 2. Show that it is not isomorphic to $(\mathbb{G}_a)^2$.

Notes.

Chapter 2 contains material on algebraic groups which can be handled with the limited amount of algebraic geometry developed in the first chapter. The material of the first three sections is fairly standard.

The Jordan decomposition in linear algebraic groups has been introduced as soon as possible, in 2.4. In our proof of the basic theorem (2.4.8) we do not follow the usual method, in which the group G is first imbedded in some $\mathbb{GL}_n$. We use, as much as possible, the formal properties of the Jordan decomposition of linear maps of a vector space.

When the Jordan decomposition is available it is not difficult to establish the basic results about commutative linear algebraic groups and diagonalizable ones. This is done in 2.4 and 2.5.

The proof of the rigidity theorem 2.5.10 shows that the affine variety V in the statement of that theorem may be replaced by any connected affine scheme over k. One thus gets a stronger result. It implies that a diagonalizable group has no "infinitesimal automorphisms".

The classification of 1-dimensional linear algebraic groups is always a tricky point in the theory. The reader may consult [11 , 7, no.4], [3 , p.257-259], [23 , no.20]. If one does not want to invoke results from curve theory (see 2.6.1), one has trouble in handling 1-dimensional unipotent groups in characteristic $p > 0$ (the geometric explanation seems to be that the affine line $\mathbb{A}^1$ is not simply connected in characteristic $p > 0$). We have given an elementary treatment of these groups, in which we need the results on polynomial cocycles contained in 2.6.5 (such results are due to M. Lazard [27, lemme 3]). We also need some results on p-polynomials. Another elementary treatment, which also requires properties of p-polynomials, can be found in [23 , no.20].

3. Derivations, differentials, Lie algebras.

3.1. Derivations and tangent spaces.

3.1.1. We first recall the definition of a derivation. Let R be a commutative ring, A a commutative R-algebra and M a left A-module. An R-*derivation* of A in M is an R-linear map $D: A \to M$ such that

$$D(ab) = a.Db + b.Da.$$

It is immediate from the definitions that $D1 = 0$, whence $D(r.1) = 0$ for all $r \in R$.
These derivations form an A-module $\mathrm{Der}_R(A,M)$, the module structure being defined by $(D+D')a = Da + D'a$ and $(bD)a = b(Da)$, if $D,D' \in \mathrm{Der}_R(A,M)$, $a,b \in A$.
The elements of $\mathrm{Der}_R(A,A)$ are called derivations of (the R-algebra) A. If $\phi: A \to B$ is a homomorphism of R-algebras, and N a B-module, then N can be viewed as an A-module in the obvious way. If $D \in \mathrm{Der}_R(B,N)$, then $D \circ \phi$ is a derivation of A in N. The map $D \mapsto D \circ \phi$ defines a homomorphism of A-modules

$$\phi_0: \mathrm{Der}_R(B,N) \to \mathrm{Der}_R(A,B),$$

whose kernel obviously is $\mathrm{Der}_A(B,N)$. Thus we obtain an exact sequence of A-modules

$$(1) \quad 0 \to \mathrm{Der}_A(B,N) \to \mathrm{Der}_R(B,N) \to \mathrm{Der}_R(A,N).$$

3.1.2. *Tangent spaces, heuristic introduction.*
Let X be a closed subvariety of $\mathbb{A}^n$. We identify its algebra

of regular functions k[X] with $k[T_1,\dots,T_n]/I$, where I is the ideal of polynomial functions vanishing on X. Assume that I is generated by $f_1,\dots,f_s$.

Let $x \in X$, let L be a line in $\mathbb{A}^n$ through x. Its points can be written as x+tv, where v is a direction vector, and t runs through k. The points of the intersection $L \cap X$ are found by solving the set of equations (for t)

(2) $f_i(x+tv) = 0,\ 1 \leqslant i \leqslant s.$

One solution is t = 0.

Let D_i be partial derivation in k[T] with respect to T_i. Then

$$f_i(x+tv) = t\sum_{j=1}^{n} a_j(D_jf_i)(x) + t^2(\dots),$$

and we see that t = 0 will be a "multiple root" of the set of equations (2) if and only if

$$\sum_{j=1}^{n} v_j(D_jf_i)(x) = 0,\ 1 \leqslant i \leqslant s.$$

If this is so, we call L a tangent line and v a tangent vector (of X in x).

Write $D' = \sum_{j=1}^{n} v_jD_j$, this is a derivation of k[T]. The last set of equation then says that $(D'f_i)(x) = 0$ $(1 \leqslant i \leqslant s)$. Denoting by M_x the maximal ideal in $k[T_1,\dots,T_m]$ of all functions vanishing in x, it follows that $D'I \subset M_x$. The linear map $f \mapsto (D'f)(x)$ factors through I and gives a linear map $D\colon k[X] \to k = k[X]/M_x$.

Viewing k as a k[X]-module k_x, via the homomorphism $f \mapsto f(x)$, we see that D is, in fact, a k-derivation of k[X] to k_x.

Conversely, every element of $\mathrm{Der}_k(k[X],k_x)$ can be obtained in this manner from a derivation D' of $k[T_1,\dots,T_n]$ with $D'I \subset M_x$ (check this). We conclude that there is a bijection of the set of tangent vectors v (such that (2) has a "multiple root t = 0") onto $\mathrm{Der}_k(k[X],k_x)$.

3.1.3. Tangent spaces.

The heuristic introduction of 3.1.2 suggests a formal intrinsic definition of tangent spaces. This we shall now introduce. First let X be an affine variety, let $x \in X$. We define the tangent space T_xX of X at x to be the k-vector space $\mathrm{Der}_k(k[X],k_x)$ (where k_x is as in 3.1.2).

If $\phi: X \to Y$ is a morphism of affine varieties, there is a corresponding algebra homomorphism $\phi^*: k[Y] \to k[X]$ and an induced homomorphism

$$\mathrm{Der}_k(k[X],k_x) \to \mathrm{Der}_k(k[Y],k_{\phi x}),$$

i.e. a linear map of tangent spaces

$$(d\phi)_x: T_xX \to T_{\phi x}Y,$$

the differential of ϕ at x, or the tangent map at x.

Clearly $d(\phi\circ\psi)_x = (d\phi)_{\psi x} \circ (x\psi)_x$.

It is also clear that the differential of an identity morphism is an identity map and that $(d\phi)_x$ is an isomorphism if ϕ is an isomorphism.

3.1.4. Lemma. *Let ϕ be an isomorphism of X onto an affine open subvariety of Y. Then $(d\phi)_x$ is an isomorphism of T_xX onto $T_{\phi x}Y$.*

Using 1.3.6 we see that it suffices to consider the case that there is $f \in k[Y]$ such that $X = \{y \in Y | f(y) \neq 0\}$. Then ϕ is the injection $X \to Y$, so $\phi x = x$.
Now we have $k[X] = k[Y][T]/(1-fT)$ (see 1.4.6). If D is a derivation of $k[Y]$ to k_x, it extends to a derivation of $k[X]$ to k_x, sending the coset t of T to $-(Df)(x)f(x)^{-2}$. From this it readily follows that the linear map, induced by ϕ,

$$\mathrm{Der}_k(k[X],k_x) \to \mathrm{Der}_k(k[Y],k_x)$$

is an isomorphism. This proves 3.1.4.

3.1.5. The preceding result expresses the "local" character of the definition of tangent space of an affine variety.
We can now define the tangent space in a point x of an arbitrary variety X (see 1.6). Let U be an affine open neighborhood of x. We define the tangent space T_xX to be T_xU. If V is a second affine open neighborhood of x, then it follows from 3.1.4 that there is a canonical isomorphism $T_xU \tilde{\to} T_xV$ (check this), which shows that the definition of T_xX makes sense.
It also is clear how to define, given a morphism of varieties $\phi: X \to Y$, a tangent map $(d\phi)_x: T_xX \to T_{\phi x}Y$.

We can now introduce the important concept of a simple point of a variety. Some basic results involving this concept will be established later in this chapter, when we have dealt with the necessary algebraic prerequisites (see 3.2.17).
Let X be a variety. We say that $x \in X$ is a <u>simple point</u> of X, or that X is <u>smooth</u> in x or that X is <u>nonsingular</u> in x if

$\dim T_x X = \dim X$ (the dimension of X defined in 1.8.1).
X is smooth, or nonsingular if all its points are so.
In the sequel, we shall only have to consider the case of an irreducible variety X.

The next result gives another description of tangent spaces. Let again X be affine, let $x \in X$ and let $M_x \subset k[X]$ be the maximal ideal of the functions vanishing in x. If $D \in T_x X$, then clearly $D(M_x^2) = 0$. Hence D defines a linear function $\lambda(D)\colon M_x/M_x^2 \to k$.

3.1.6. Lemma. *λ is an isomorphism of $T_x X$ onto the dual of the vector space M_x/M_x^2.*
Let ℓ be a linear function on M_x/M_x^2. Define a linear map $\mu\ell\colon k[X] \to k$ by $\mu(\ell)f = \ell(f-f(x)+M_x^2)$. Then $\mu\ell \in T_x X$ and $\lambda\circ\mu = \mathrm{id}$, $\mu\circ\lambda = \mathrm{id}$. We leave the details to the reader.

3.1.7. Exercises. (1) Using 3.1.6, describe $T_x X$ in the following cases: (a) X is a point, (b) $X = \mathbb{A}^n$, (c) $X = \{(\xi_1,\xi_2) \in \mathbb{A}^2 \mid \xi_1\xi_2 = 0\}$, $x = (0,0)$, (d) (char $k \neq 2,3$) $X = \{(\xi_1,\xi_2) \in \mathbb{A}^2 \mid \xi_1^2 = \xi_2^3\}$, $x = (0,0)$.
(2) Let X be an affine variety, let $\mathcal{O}_{X,x}$ be the local ring of X in $x \in X$ (see 1.4.4(1)). Let M_x be the maximal ideal of $\mathcal{O}_{X,x}$. Then $T_x X$ is isomorphic to the dual of the vector space M_x/M_x^2.
(3) Let X and Y be varieties, let $x \in X$, $y \in Y$. Then $T_{(x,y)}(X\times Y) \simeq T_x(X) \oplus T_y(Y)$.
(4) If X is a closed subvariety of Y, and $\phi\colon X \to Y$ the injection, then $(d\phi)_x\colon T_x X \to T_x Y$ is injective.

We shall need a number of results about derivations, in particular about derivations of fields. To deal with them it seems best to introduce differentials.

3.2. *Differentials, separability.*

3.2.1. Let R be a commutative ring and A a commutative R-algebra. Denote by $q: A \otimes_R A \to A$ the product homomorphism (so $q(a \otimes b) = ab$) and let $I = \operatorname{Ker} q$. This ideal of $A \otimes_R A$ is generated by the elements $a \otimes 1 - 1 \otimes a$ ($a \in A$). The quotient algebra $A \otimes_R A/I$ is isomorphic to A. We define the *module of differentials* $\Omega_{A/R}$ (of the R-algebra A) by

$$\Omega_{A/R} = I/I^2.$$

This is an $A \otimes_R A$-module, but since $I\Omega_{A/R} = 0$, we may - and shall - view it as an A-module.

Denote by da (or $d_{A/R}a$, if necessary) the image of $a \otimes 1 - 1 \otimes a$ in $\Omega_{A/R}$. This is the differential of a. One checks immediately that d is a derivation of A in $\Omega_{A/R}$, and that the da generate the A-module $\Omega_{A/R}$.

The basic property of $\Omega_{A/R}$, expressed by the following result, is that it is a "universal module for derivations".

3.2.2. *Theorem.* (i) *For every A-module M, the map* $\operatorname{Hom}_A(\Omega_{A/R}, M) \to \operatorname{Der}_R(A, M)$ *with* $\phi \mapsto \phi \circ d$ *is an isomorphism of A-modules;*

(ii) *The pair* $(\Omega_{A/R}, d)$ *of an A-module, together with an R-derivation d of A in* $\Omega_{A/R}$, *is determined up to an isomorphism of A-modules by the property of* (i).

Let Φ be map of (i). It is obvious that Φ is a homomorphism of A-modules. Since the da generate the A-module $\Omega_{A/R}$ we have that Φ is injective.
Next let $D \in \mathrm{Der}_R(A,M)$. Define an R-linear map $\psi\colon A \otimes_R A \to M$ by $\psi(a \otimes b) = bDa$. Then

$$\psi(xy) = q(x)\psi(y) + q(y)\psi(x).$$

It follows that ψ vanishes on I^2, hence ψ defines an R-linear map $\phi\colon \Omega_{A/R} \to M$, which in fact is A-linear. Then $\psi(a \otimes 1 - 1 \otimes a) = da$, whence $\Phi\phi = \phi \circ d = D$, proving that Φ is surjective. This proves (i).
The proof of (ii) is standard.

3.2.3. If $\phi\colon A \to B$ is a homomorphism of R-algebras, the definitions of 3.2.1 show that there is a unique homomorphism of A-modules

$$\phi^0\colon \Omega_{A/R} \to \Omega_{B/R},$$

with $\phi^0 \circ d_{A/R} = d_{B/R} \circ \phi$.
If N is a B-module and if ϕ_0 is as in 3.1.1, there is a commutative diagram of A-modules

$$\begin{array}{ccc} \mathrm{Der}_R(B,N) & \tilde{\to} & \mathrm{Hom}_B(\Omega_{B/R},N) \\ {\scriptstyle\phi_0}\downarrow & & \downarrow \\ \mathrm{Der}_R(A,N) & \tilde{\to} & \mathrm{Hom}_A(\Omega_{A/R},N), \end{array}$$

where the horizontal isomorphisms are as in 3.2.2, and the right-hand arrow is induced by ϕ^0. This is easily checked.

Now let A be an R-algebra of the form

$$A = R[T_1,\ldots,T_n]/(f_1,\ldots,f_m).$$

Let t_i be the image in A of T_i, put $t = (t_1,\ldots,t_n)$. Denote by D_i partial derivation in $R[T_1,\ldots,T_n]$ with respect to T_i $(1 \leq i \leq n)$.

3.2.4. _Lemma_. _The_ dt_i $(1 \leq i \leq n)$ _generate the_ A-_module_ $\Omega_{A/R}$. _The kernel of the homomorphism_ $\phi: A^n \to \Omega_{A/R}$ _with_ $\phi e_i = dt_i$ _is the_ A-_submodule generated by the elements_ $\sum_{i=1}^{n} (D_i f_j)(t) e_i$ $(1 \leq j \leq m)$.

Here (e_i) is the standard basis of A^n.

Let D be an R-derivation of A in an A-module M. If $f \in R[T_1,\ldots,T_n]$ then

$$D(f(t)) = \sum_{i=1}^{n} (D_i f)(t).Dt_i,$$

whence

$$\sum_{i=1}^{n} (D_i f_j)(t).Dt_i = 0 \quad (1 \leq j \leq m).$$

Let K be the submodule of A^n described in the lemma. One now readily checks that the A-module A^n/K has the universal property of 3.2.2, hence is isomorphic to $\Omega_{A/R}$, by 3.2.2(ii). The assertion follows.

3.2.5. _Exercises_. (1) If $A = R[T_1,\ldots,T_n]$ then $\Omega_{A/R}$ is a free A-module with basis $(dT_i)_{1 \leq i \leq n}$.

(2) In the situation of 3.2.4, with $m = n = 1$, give a condition on $f = f_1$ under which $\Omega_{A/R} = 0$. Discuss the particular case that R is a field.

(3) Let A be a R-algebra which is an integral domain, let F

be the quotient field of A. Then $\Omega_{F/R} \simeq \Omega_{A/R} \otimes_A F$.

(4) Let F be a field, let $E = F(x_1,\ldots,x_n)$ be an extension field of finite type. Then $\Omega_{E/F}$ is a finite dimensional vector space over F, spanned by the dx_i.

(5) Let $A = k[T_1,T_2]/(T_1^2-T_2^3)$. Show that $\Omega_{A/k}$ is not a free A-module.

(6) Let A and B be R-algebras, and $A \otimes_R B$ their tensor product. There is an isomorphism of $A \otimes_R B$-modules $\Omega_{A\otimes_R B/R} \xrightarrow{\sim} (\Omega_{A/R} \otimes_R B) \oplus (A \otimes_R \Omega_{B/R})$ under which $d_{A\otimes_R B/R}$ corresponds to $(d_{A/R} \otimes id_B) \oplus (id_A \otimes d_{B/R})$.

3.2.6. We next discuss in more detail the case of fields. Let F be a field and E,E' two extensions of F of finite type, with $E' \subset E$.

By (3.1(1)) we have an exact sequence of vector spaces over E

$$0 \to Der_{E'}(E,E) \to Der_F(E,E) \to Der_F(E',E),$$

whence, by 3.2.2(i), an exact sequence of E-vector spaces

$$0 \to Hom_E(\Omega_{E/E'},E) \to Hom_E(\Omega_{E/F},E) \to Hom_{E'}(\Omega_{E'/F},E).$$

Since the map $u \mapsto u \otimes 1$ of $\Omega_{E'/F}$ into $\Omega_{E'/F} \otimes_{E'} E$ defines an isomorphism of vector spaces over E

$$Hom_{E'}(\Omega_{E'/F} \otimes_{E'} E,E) \xrightarrow{\sim} Hom_{E'}(\Omega_{E'/F},E)$$

we get an exact sequence of vector spaces over E

$$0 \to Hom_E(\Omega_{E'/E'},E) \to Hom_E(\Omega_{E/F},E) \to Hom_E(\Omega_{E'/F} \otimes_{E'} E,E),$$

and all vector spaces involved are finite dimensional (by

3.2.5(4)). Dualizing we obtain an exact sequence of vector spaces over E'

$$(1) \quad \Omega_{E'/F} \otimes_{E'} E \overset{\alpha}{\to} \Omega_{E/F} \to \Omega_{E/E'} \to 0,$$

which is basic in the following discussion. Notice that $\alpha(d_{E'/F}a \otimes 1) = d_{E/F}a \quad (a \in E')$.

Recall that E is said to be <u>separably</u> <u>algebraic</u> over F if for each $x \in E$ there is a polynomial $f \in F[T]$ without multiple roots such that $f(x) = 0$ [26 , Ch. VII, §4]. We may assume f to be irreducible and in that case its derivative f' is nonzero. If char F = 0 then every algebraic extension of F is separable.

3.2.7. <u>Lemma</u>. <u>If</u> E <u>is</u> <u>separably</u> <u>algebraic</u> <u>over</u> E' <u>then</u> <u>the</u> <u>homomorphism</u> α <u>is</u> <u>injective</u>.

From the discussion in 3.2.6 we see that injectivity of α is equivalent to the surjectivity of the homomorphism of 3.1.1

$$\mathrm{Der}_F(E,E) \to \mathrm{Der}_F(E',E).$$

An equivalent property is: any F-derivative of E' in E can be extended to a F-derivation of E in E. To prove this, it suffices to deal with the case of a simple extension $E = E'(x) \cong E'[T]/(f)$, where f is an irreducible polynomial with $f'(x) \neq 0$. Let $D \in \mathrm{Der}_F(E',E)$. If $g = \sum_{i \geq 0} a_i T^i \in E[T]$, put $Dg = \sum_{i \geq 0} (Da_i)T^i$, so $Dg \in E[T]$.

Then (check this) D is extendable to an F-derivation D' of E in E with $D'x = a$ if and only if

$f'(x)a + (Df)(x) = 0.$

Since $f'(x) \neq 0$, this equation has a unique solution, establishing what we wanted.

3.2.8. Lemma. Let $E = F(x)$. Then $\dim_E \Omega_{E/F} \leqslant 1$. *We have* $\Omega_{E/F} = 0$ *if and only if* x *is separably algebraic over* F.
If x is transcendental over F this is a consequence of 3.2.5 (1). If x is algebraic over F we have $E \simeq F[T]/(f)$, where f is the minimum polynomial of x over F. The assertion then follows from 3.2.4, using that $f'(x) \neq 0$ if and only if x is separably algebraic.

Let, as before, E be an extension of finite type F. We denote by $\text{trdeg}_F E$ the transcendence degree of E over F. If $E = F(x_1,\ldots,x_r)$, the transcendence degree t is the maximal number of x_i which are algebraically independent over F. If $r = t$ then E is purely transcendental over F. We say that E is *separably generated* over F if it is a separable algebraic extension of a purely transcendental extension of F. Let $p = \text{char } F$. If $p = 0$ then E is always separably generated. Recall that if $p > 0$ the field F is said to be *perfect* if $x \mapsto x^p$ is a surjective map of F onto F. If $p = 0$, then F is always perfect. We can now state the main result about fields.

3.2.9. Theorem. (i) $\dim_E \Omega_{E/F} \geqslant \text{trdeg}_F E$;
(ii) *Equality holds in* (i) *if and only if* E *is separably generated over* F;
(iii) *If* F *is perfect then* E *is separably generated over* F.

We prove (i) and (ii) together, by induction on $t = \mathrm{trdeg}_F E$. The following is used (see 3.2.6): if $x \in E-F$ there is an exact sequence

$$(2) \quad \Omega_{F(x)/F} \otimes_F E \overset{\alpha}{\to} \Omega_{E/F} \to \Omega_{E/F(x)} \to 0,$$

and $\alpha(d_{F(x)/F}\, a \otimes 1) = d_{E/F} a$.

First let $t = 0$. Then (i) is trivial. Now let $\Omega_{E/F} = 0$. Assume $E = F(x_1,\dots,x_r)$. If $r = 1$, lemma 3.2.8 shows that E is separably algebraic over F. If $r > 1$, then (2), with $x = x_1$, shows that $\Omega_{E/F(x_1)} = 0$. By an induction on r we then may conclude that E is separably algebraic over $F(x_1)$. But then 3.2.7 implies that $\Omega_{F(x_1)/F} = 0$, and by 3.2.8 we have that x_1 is separably algebraic over F. This establishes that E is a separable algebraic extension of F. Now let $t > 0$, and assume (i) and (ii) true for transcendence degree t-1.

By what we have already proved, there is $x \in E$ with $d_{E/F} x \neq 0$. Apply (2) with this x. Since $\alpha(d_{F(x)/F} x \otimes 1) = d_{E/F} x \neq 0$, we have $\Omega_{F(x)/F} \neq 0$. Since $\dim \Omega_{F(x)/F} \leqslant 1$ by 3.2.8, we must have that α is injective and $\dim \Omega_{F(x)/F} = 1$. Consequently

$\dim \Omega_{E/F} = \dim \Omega_{E/F(x)} + 1$.

Since $\mathrm{trdeg}_F E = \mathrm{trdeg}_{F(x)} E + \mathrm{trdeg}_F F(x)$ (see e.g. [26 , Ch. X, 8.5]) and $\mathrm{trdeg}_F F(x) \leqslant 1$ we have $\mathrm{trdeg}_{F(x)} E \geqslant t - 1$ and it follows that $\dim \Omega_{E/F} \geqslant t$, which establishes (i).

If $\dim \Omega_{E/F} = t$, then we must have $\dim \Omega_{E/F(x)} = t-1$.

It follows from (i) and the preceding remarks that

$\dim \Omega_{E/F(x)} = \mathrm{trdeg}_{F(x)} E = t-1$. The induction assumption for (ii) shows that E is a separable algebraic extension of a

purely transcendental extension of F(x). Since $\mathrm{trdeg}_{F(x)}E = t-1$, we must have that x is transcendental over F, and it follows that E is separably generated over F.
To finish the proof of (ii) we have to show that $\dim \Omega_{E/F} = t$ if E is separably generated over F. If t = 0, then E is separably algebraic over F. Using (4) one establishes, by an induction on the number of generators, using 3.2.7 and 3.2.8, that $\Omega_{E/F} = 0 = t$. If $t > 0$ apply (2) with $E' = F(x_1,\dots,x_t)$ a purely transcendental extension of F such that E is separably algebraic over E'. Then $\Omega_{E/E'} = 0$ by what we already proved and 3.2.7 shows that α is injective. It follows from 3.2.5(1), (3) that $\dim_{E'}\Omega_{E'/F} = t$, from which one readily concludes that $\dim \Omega_{E/F} = t$. This finishes the proof of (ii).
To prove (iii) we use induction on the number of generators of $E = F(x_1,\dots,x_r)$, starting with r = 0. We assume that $p = \mathrm{char}\, F > 0$. If t = 0 then (iii) is true, because an algebraic extension of a perfect field is separable (see [26 , Ch. VII, §7]). Assume $t > 0$. If r = t then (iii) is obvious. So let $t < r$, and assume that $x_1,\dots,x_t$ are algebraically independent over F. If $t < r-1$ we may by induction assume that there are algebraically independent elements $y_1,\dots,y_t$ in $F(x_1,\dots,x_{r-1})$ such that $F(x_1,\dots,x_{r-1})$ is separably algebraic over $F(y_1,\dots,y_t)$. Likewise, there are algebraically independent elements $z_1,\dots,z_t \in F(y_1,\dots,y_t,x_r)$ such that $F(y_1,\dots,y_t,x_z)$ is separably algebraic over $F(z_1,\dots,z_t)$. It then follows, using a transitivity property of separably algebraic extensions (see [26 , Ch. VII, §4]) that E is separa-

bly algebraic over $F(z_1,\ldots,z_t)$.
This reduces the proof of (iii) to the case $t = r-1$. From the fact that F is perfect it follows that there is a nonzero polynomial $f \in F[T_1,\ldots,T_r]$ such that $f(x_1,\ldots,x_r) = 0$ and that not all exponents of the powers of the indeterminates occurring in f are divisible by p. Using 3.2.4, applied with $R = F$ and $A = F[x_1,\ldots,x_r]$, and 3.2.5(3) it follows that $\dim \Omega_{E/F} \leqslant r-1 = t$ (notice that in this case the kernel of the homomorphism ϕ of 3.2.4 is nonzero). Now (iii) follows from (i) and (ii).

Let E,E',F be as in 3.2.6.

3.2.10. Corollary. *Assume F to be perfect. Either of the following conditions is necessary and sufficient for E to be separably generated over E':*

(a) $\alpha\colon \Omega_{E'/F} \otimes_{E'} E \to \Omega_{E/F}$ *is injective;*

(b) $\mathrm{Der}_F(E,E) \to \mathrm{Der}_F(E',E)$ *is surjective.*

The equivalence of (a) and (b) follows from 3.2.6. To establish the assertion about (a), use the exact sequence (1) of 3.2.6, and observe that by 3.2.9(ii), (iii) we have $\dim \Omega_{E'/F} = \mathrm{trdeg}_F E'$, $\dim \Omega_{E/F} = \mathrm{trdeg}_F E$. It follows that $\dim \Omega_{E/E'} = \mathrm{trdeg}_{E'} E$ if and only if α is injective. Using 3.2.9(ii), we get 3.2.10.

3.2.11. Exercises. (1) Let E,E' and F be as in 3.2.10. If E is separably algebraic over E' then α is an isomorphism.
(2) Let F be perfect, let E be an extension of F of finite type. Put $p = \mathrm{char}\, F$. If $a \in E$, $d_{E/F}a = 0$ then $a \in F$ if $p = 0$

and a is a pth power of an element of E if $p > 0$ (Hint: use 3.2.9(iii) and (3.1(1)).

3.2.12. We are going to apply the preceding results to some geometric questions. First some preparations.

Let R be an integral domain, with quotient field F. If $f \in R$, $f \neq 0$, denote by R_f the ring $R[T]/(1-fT)$ (see also 1.4.6).

We may and shall view this as the subring $R[f^{-1}]$ of F, i.e. the ring of all elements of F of the form $f^{-n}a$ ($a \in R$, $n \geq 0$).

If M is an R-module we denote by M_f the R_f-module $M \otimes_R R_f$.

Let $A = (a_{ij})_{\substack{1 \leq i \leq m \\ 1 \leq j \leq n}}$ be an m×n-matrix with entries in R.

Denote by r the rank of A, viewed as a matrix with entries in F.

Define the R-module M(A) (or $M_R(A)$) to be the quotient of R^n by the submodule generated by the elements $\sum_{j=1}^{n} a_{ij}e_j$ $(1 \leq i \leq m)$, (e_j) being the canonical basis.

We denote by $GL_n(R)$ the group of those n×n-matrix with entries in R, which have an inverse with entries in R.

3.2.13. Lemma. Let A be as above.

(i) If $B \in GL_m(R)$ then $M(BA) = M(A)$, if $C \in GL_n(R)$ then $M(AC) \simeq M(A)$;

(ii) There exists $f \in R$, $f \neq 0$ and $B \in GL_m(R_f)$, $C \in GL_n(R_f)$ such that

$$A = B \begin{pmatrix} I_r & 0 \\ 0 & 0 \end{pmatrix} C.$$

Here I_r is the r×r-identity matrix. (i) is easy.

The assertion of (ii) is true if R is a field, by elementary linear algebra (view A as the matrix of a linear map $R^n \to R^m$ and choose suitable bases). In the general case, take $B \in GL_m(F)$, $C \in GL_n(F)$ with the required property and choose f such that B, C, B^{-1}, C^{-1} have entries in R_f.

3.2.14. <u>Lemma</u>. <u>There is</u> $f \in R$, $f \neq 0$ <u>such that</u> $M(A)_f$ <u>is a free</u> R_f<u>-module of rank</u> $n-r$. <u>We may choose</u> f <u>such that</u> $n-r$ <u>of the images</u> e_i' <u>of the elements</u> $e_i \otimes 1$ <u>of</u> $R^n \otimes R_f$ <u>form a basis of</u> $M(A)_f$.

The first point follows from 3.2.13. Let $(f_1, \dots, f_{nr})$ be a basis of $M(A)_f$ and assume that the elements $e_{r+1}', \dots, e_n'$ are linearly independent. We can write

$$e_{r+i}' = \sum_{j=1}^{n-r} c_{ij} f_j \quad (1 \leq i \leq n-r),$$

where $c_{ij} \in R_f$, $\det(c_{ij}) \neq 0$. By modifying f, if necessary, we may assume that the inverse matrix $(c_{ij})^{-1}$ has entries in R_f. Then $e_{r+1}', \dots, e_n'$ are as required.

3.2.15. Now let X be an irreducible affine variety over k. We are going to apply the preceding results for $R = k[X], F = k(X)$. Recall that the field k is algebraically closed, hence perfect. If $x \in X$, let $M_x = \{f \in k[X] \ f(x) = 0\}$ be the maximal ideal defined by x. If M is a k[X]-module, put

$$M(x) = M/M_x M,$$

this is a vector space over k (since $M/M_x \simeq k$).

Let A be as in 3.2.12, with $R = k[X]$. Since the entries of A are functions on X, the matrix A(x) with entries in k is defined. It is immediate that

$$M_{k[X]}(A)(x) = M_k(A(x)).$$

3.2.16. _Lemma_. _Let_ r _be as before_.
(i) _The set of_ $x \in X$ _such that_ $\dim_k M_{k[X]}(A)(x) = n-r$ _is open and non-empty_;
(ii) _If_ $x \in X$ _and_ $\dim M_{k[X]}(A)(x) = n-r$ _there is_ $f \in k[X]$ _with_ $f(x) \neq 0$ _such that_ $M_{k[X]}(A)_f$ _is a free_ $k[X]_f$_-module of rank_ $n-r$;
(iii) $\dim_{k(X)} M_{k[X]}(A) \otimes_{k[X]} k(X) = n-r$.

By familiar results from linear algebra, we have that $\dim M_{k[X]}(A)(x) = n-r$ if and only if r is the maximal size of a square submatrix of A(x) with nonzero determinant. (i) now follows readily from the fact that the rank of A over k(X) equals r.

Let x be as in (ii). We may assume that $\det(a_{ij}(x))_{1\leq i,j\leq r} \neq 0$. Put $f = \det(a_{ij})_{1\leq i,j\leq r}$. If $e_i' \in M(A)_f$ is as in 3.2.14, then

$$\sum_{j=1}^{n} a_{ij}e_j' = 0 \quad (1 \leq i \leq m).$$

Our assumption implies that we can express e_i' $(1 \leq i \leq r)$ as a linear combination, with coefficients in $k[X]_f$, of $e_{r+1}',\dots,e_n'$. It also implies that the elements $e_{r+1}' \otimes 1,\dots,e_n' \otimes 1$ of the vector space $M(A)_f \otimes_{k(X)_f} k(X)$ form a basis, which shows that $e_{r+1}',\dots,e_n'$ are linear independent over $k[X]_f$. Hence $(e_{r+1}',\dots,e_n')$ is a basis of $M(A)_f$, establishing (ii) and also (iii).

We put $\Omega_X = \Omega_{k[X]/k}$. If $x \in X$, the tangent space $T_xX = \mathrm{Der}_k(k[X],k_x)$ (see 3.1.3) is, by 3.2.2(i), isomorphic to $\mathrm{Hom}_{k[X]}(\Omega_X,k_x)$. Since for any $k[X]$-module M we have $\mathrm{Hom}_{k[X]}(M,k_x) \simeq \mathrm{Hom}_k(M(x),k)$ (check this), it follows that

$$T_xX \simeq \mathrm{Hom}_k(\Omega_X(x),k).$$

The dual vector space of the tangent space T_xX is called the <u>cotangent space</u> $(T_xX)^*$ (of X at x). It can be identified with the vector space $\Omega_X(x)$.

We can now give some basic results on simple points (see 3.1.5).

3.2.17. <u>Theorem</u>. <u>Let</u> X <u>be an irreducible variety of dimension</u> e.

(i) <u>If</u> x <u>is a simple point of</u> X, <u>there is an affine open neighborhood</u> U of X <u>such that</u> Ω_U <u>is a free</u> $k[X]$-<u>module with a basis</u> $(df_1,\dots,df_e)$, <u>for suitable</u> $f_i \in k[U]$;

(ii) <u>The simple points of</u> X <u>form a non-empty open subset of</u> X.

We may assume that X is affine.

Let $k[X] = k[T_1,\dots,T_n]/(f_1,\dots,f_m)$. By 3.2.4, we have $\Omega_X \simeq M_{k[X]}(A)$, where $A = (D_jf_i)_{\substack{1\leqslant j\leqslant m\\ 1\leqslant j\leqslant n}}$.

From 3.2.9(ii),(iii) we see that $\dim_{k(X)}\Omega_{k(x)/k} = e$. If r is the rank of A (as a matrix over $k(X)$), then 3.2.16(iii) shows that $e = n-r$. Now (i) follows from 3.2.16(ii) and 3.2.14 and (ii) is a consequence of 3.2.16(i).

If $f_1,\dots,f_e$ are as in 3.2.17(i) we say that they are a <u>system of local coordinates</u> at the simple point x (There are good reasons for this terminology, see for example [28 , p. 361-362]. We shall not go into these here).

3.2.18. <u>Exercises</u>. (1) The elements of a system of local coordinates are algebraically independent.
(2) Let X be an irreducible affine variety. If Ω_X is a free k[X]-module then X is smooth.

3.2.19. Let $\phi: X \to Y$ be a morphism of irreducible varieties. It is called <u>dominant</u> if ϕX is dense in Y. In that case it follows from 1.9.1(ii) that there is an injection ϕ^* of quotient fields $k(Y) \to k(X)$. So we can view k(X) as an extension of k(Y). We say that ϕ is <u>separable</u> if this extension is separably generated.
Let, moreover, X and Y be affine and denote by ϕ^* also the injective homomorphism $k[Y] \to k[X]$ defined by ϕ. According to 3.2.3 we have a homomorphism of k[Y]-modules

$$(\phi^*)^0 : \Omega_Y \to \Omega_X .$$

Let $x \in X$, let k_x be as in 3.1.2. Viewed as a k[Y]-module (via ϕ^*) it is $k_{\phi x}$ (check this).
We are going to study the linear map of 3.1.3

$$(d\phi)_x : T_x X \to T_{\phi x} Y .$$

Using the preceding remarks and the diagram of 3.2.3, we see that we can view $(d\phi)_x$ also as the homomorphism

$$\mathrm{Hom}_{k[X]}(\Omega_X,k_x) \to \mathrm{Hom}_{k[Y]}(\Omega_Y,k_{\phi x})$$

deduced from $(\phi^*)^0$. It can also be viewed as a linear map

$$\mathrm{Hom}_k(\Omega_X(x),k) \to \mathrm{Hom}_k(\Omega_Y(\phi x),k).$$

3.2.20. Theorem. Let X and Y be irreducible varieties and $\phi: X \to Y$ a morphism.
(i) Assume that $x \in X$ and $\phi x \in Y$ are simple points of X and Y, respectively, such that $(d\phi)_x: T_xX \to T_{\phi x}Y$ is surjective. Then ϕ is dominant and separable;
(ii) Assume that ϕ is dominant and separable. There is a non-empty open subset U of X such that for $x \in U$, the point ϕx is simple on Y and that $(d\phi)_x: T_xX \to T_{\phi x}Y$ is surjective.
Using 3.2.17 we see that it suffices to consider the case that X and Y are affine and smooth, and that Ω_X and Ω_Y are free modules over k[X],k[Y], respectively, whose ranks are $m = \dim X$ and $n = \dim Y$.
The homomorphism $(\phi^*)^0$ of 3.2.19 leads to a homomorphism of free k[X]-modules

$$\psi: \Omega_Y \otimes_{k[Y]} k[X] \to \Omega_X.$$

Fixing bases of these modules, ψ is described by an m×n-matrix $A = (a_{ij})_{\substack{1 \leq i \leq m \\ 1 \leq i \leq n}}$ with entries in k[X].

Now if $x \in X$ is such that $(d\phi)_x$ is surjective, the remarks of 3.2.19 imply that the matrix A(x) has rank n.
An argument involving determinants, as in the proof of

3.2.16(i), then shows that the rank A is at least n. Since it is obviously at most n, this rank equals n.
Consequently ψ is injective. Then the homomorphism $(\phi^*)^0$ of 3.2.19 is also injective. Since Ω_X and Ω_Y are free modules over k[X] and k[Y], it follows that the homomorphism ϕ^*: k[Y] → k[X] is injective. This means that ϕ is dominant (1.9.1(ii)). Using 3.2.5(3) it also follows that the homomorphism α of 3.2.6

$$\Omega_{k(Y)/k} \otimes_{k(Y)} k(X) \to \Omega_{k(X)/k}$$

is injective (in fact, on appropriate bases, it is given by the matrix A).
The separability of ϕ now follows from 3.2.10. This establishes (i).
Conversely, if ϕ is dominant and separable the rank of A, as a matrix with entries on k(X), is n. Hence there is a non-empty open set of points $x \in X$ such that the rank of A(x) equals n, whence (ii) (using 3.2.17(ii)).

To conclude this section we give some consequences of the preceding results for homogeneous spaces of algebraic groups (see 2.3.1).

3.2.21. *Theorem. Let G be a connected algebraic group.*
(i) Let X be a homogeneous space for G. Then X is irreducible and smooth. In particular, G is smooth;
(ii) Let ϕ: X → Y be a G-morphism of homogeneous spaces. Then ϕ is separable if and only if the tangent map $(d\phi)_x$ is sur-

jective for some $x \in X$. _If this is so, then_ $(d\phi)_x$ _is surjective for all_ $x \in X$;

(iii) _Let_ $\phi: G \to G'$ _be a surjective homomorphism of algebraic groups. Then_ ϕ _is separable if and only if_ $(d\phi)_e$ _is surjective._

If X is a homogeneous space and $x \in X$, the morphism $G \to X$ sending g to g.x. is surjective, hence X is irreducible (1.2.3(ii)). If $g \in G$, the map $x \mapsto g.x$. is an isomorphism of the variety X onto itself. Hence x is simple if and only if g.x is simple.

(i) now follows from 3.2.17(ii) and (ii) from 3.2.20. Finally, (iii) is a consequence of (ii), viewing G and G' as homogeneous spaces for G.

3.3. The Lie algebra of a linear algebraic group.

3.3.1. Let G be a linear algebraic group. Denote by λ and ρ left- and right translations in $A = k[G]$ (2.3). Let $q: A \otimes_k A \to A$ be the product homomorphism. If $I = \operatorname{Ker} q$ then $\Omega_G = I/I^2$ (3.2.1). Viewing $A \otimes_k A$ as the algebra $k[G \times G]$ of regular functions on $G \times G$ we have $(qF)(x) = F(x,x)$ ($F \in k[G \times G]$). If $x \in G$ we denote by $\lambda(x)$ and $\rho(x)$ the automorphisms of $k[G \times G]$ defined by

$$(\lambda(x)F)(y,z) = F(x^{-1}y,x^{-1}z),(\rho(x)F)(y,z) = F(yx,zx).$$

It is clear that all $\lambda(x)$ and $\rho(x)$ stabilize I, hence induce automorphisms of Ω_G, also denoted by $\lambda(x)$ and $\rho(x)$.

We thus have defined representations λ and ρ of G in Ω_G, which are locally finite (by 2.3.4(i)). It is clear that the derivation d: $A \to \Omega_G$ (3.2.1) commutes with all $\lambda(x)$ and $\rho(x)$ ($x \in G$). If $x \in G$, then Int x: $y \mapsto xyx^{-1}$ is an automorphism of the algebraic group G. Since it fixes e, it induces automorphisms Ad x of the tangent space T_eG and $(\mathrm{Ad}x)^*$ of the cotangent space $(T_eG)^* = \Omega_G(e)$. Thus, if u is an element of $(T_eG)^*$, i.e. a linear function on T_eG, we have

$$((\mathrm{Ad}\ x)^*u)X = u(\mathrm{Ad}(x^{-1})X) \quad (x \in G, X \in T_eG).$$

Let $M_e \subset A$ be the maximal ideal of functions vanishing in e. Then $(T_eG)^*$ can be identified with M_e/M_e^2 (3.1.6). If $f \in A$ we denote by $\delta f \in (T_eG)^*$ the element $f-f(e)+M_e^2$. Then $\delta f(X) = X(f-f(e)) = Xf$, by the proof of 3.1.6.

3.3.2. Proposition. *There is an isomorphism of* k[G]-*modules*

$$\Phi: \Omega_G \xrightarrow{\sim} k[G] \otimes_k (T_eG)^*,$$

the module structure of the righthand side being defined by

$$f(g \otimes u) = (fg) \otimes u \quad (f,g \in k[G], u \in (T_eG)^*).$$

We may require Φ *to have the following properties:*

(a) $\Phi \circ \lambda(x) \circ \Phi^{-1} = \lambda(x) \otimes \mathrm{id}$, $\Phi \circ \rho(x) \circ \Phi^{-1} = \rho(x) \otimes (\mathrm{Ad}\ x)^*$ ($x \in G$);

(b) *Let* d *be the derivation* $A \to \Omega_G$. *If* $f \in A$ *and* $f(x,y) = \sum_{i=1}^{r} f_i(x)g_i(y)$ *with* $f_i, g_i \in A$, *then*

$$(\Phi \circ d \circ \Phi^{-1})f = -\sum_{i=1}^{r} f_i \otimes \delta g_i.$$

The map $(x,y) \mapsto (x,xy)$ is an isomorphism of the algebraic variety $G \times G$. It induces an isomorphism ψ of the algebra

$k[G \times G] = k[G] \otimes k[G]$ such that

$$(\psi F)(x,y) = F(x,xy),$$

$$\psi \circ (\lambda(x) \otimes \mathrm{id}) = (\lambda(x) \otimes \lambda(x)) \circ \psi,$$

$$\psi \circ (\rho(x) \otimes \mathrm{Int}\ x) = (\rho(x) \otimes \rho(x)) \circ \psi.$$

It follows that ψI is the ideal of functions vanishing on $G \times \{e\}$, which is $k[G] \otimes_k M_e$. Then $\psi I^2 = k[G] \otimes_k M_e^2$, whence $\Omega_G \simeq k[G] \otimes_k (M_e/M_e^2)$. This proves the first assertion (using 3.1.6).

The proofs of the assertion about the module structure and of (a) are straightforward and can be left to the reader. The proof of (b) follows from

$$\psi(f \otimes 1 - 1 \otimes f)(x,y) = \sum_{i=1}^{r} f_i(x)(g_i(e) - g_i(y))$$

(notations as in (b)).

3.3.3. We assume that the reader is familiar with the basic facts about Lie algebras (which can be found in [22] or [24]).

If A is a k-algebra, then the space $\mathcal{D} = \mathrm{Der}_k(A,A)$ is a Lie algebra: if D_1, D_2 are derivations of A (in A) then $[D_1,D_2] = D_1 \circ D_2 - D_2 \circ D_1$ is also one, as is readily checked, and $(D_1,D_2) \mapsto [D_1,D_2]$ defines on $\mathcal{D}$ a Lie algebra structure.

If $p = \mathrm{char}\ k > 0$ then Leibniz' formula shows that

$$D^p(ab) = \sum_{i=0}^{p} \binom{p}{i} a^i b^{p-i} = a.D^p b + b.D^p a \quad (a,b \in A),$$

so that D^p is again a derivation. In this case $\mathcal{D}$ is a <u>restricted Lie algebra</u> (or a p-Lie algebra). This means that $\mathcal{D}$ has a "p-operation" $D \mapsto D^{[p]}$ (which in our case is the ordinary p^{th} power) such that the following holds (we put $(\text{ad } D)D' = [D,D']$):
(a) $(\xi D)^{[p]} = \xi^p D^{[p]}$ $(\xi \in k)$, (b) $(\text{ad } D)^p = \text{ad}(D^{[p]})$,
(c) $(D+D')^{[p]} = D^{[p]} + D'^{[p]} + \sum_{i=1}^{p-1} i^{-1} s_i(D,D')$,
where $s_i(D,D')$ is the coefficient of ξ^i in $\text{ad}(\xi D+D')^{p-1}(D)$ ("Jacobson's formula").
We shall not need the formulas of (b) and (c). For a further discussion see [8 , Ch. I, p.105-106] or [24 , Ch. V, no.7].
Now let G and A be as in 3.3.1. If necessary, we write $\mathcal{D} = \mathcal{D}_G$. Then λ and ρ define representations of G in $\mathcal{D}$, denoted by the same symbols. So $\lambda(x)D = \lambda(x) \circ D \circ \lambda(x)^{-1}$, $\rho(x)D = \rho(x) \circ D \circ \rho(x)^{-1}$ $(x \in G, D \in \mathcal{D})$. The <u>Lie algebra</u> L(G) (or Lie(G)) of G is the set of $D \in \mathcal{D}$ commuting with all $\lambda(x)$ $(x \in G)$. It is immediate that L(G) is a subalgebra of the Lie algebra $\mathcal{D}$, stable under the p-operation if $p > 0$. Since $\lambda(x)$ and $\rho(y)$ commute $(x,y \in G)$ we have that all $\rho(y)$ stabilize L(G). We denote the induced linear maps also by $\rho(y)$.

3.3.4. <u>Corollary</u>. <u>There is an isomorphism of</u> k[G]-<u>modules</u> $\Psi: \mathcal{D}_G \xrightarrow{\sim} k[G] \otimes_k T_e G$, <u>the module structure of the right-hand side being defined by</u>

$$f(g \otimes X) = (fg) \otimes X \quad (f,g \in k[G], X \in T_e G).$$

<u>We may require</u> Ψ <u>to have the following properties</u>
(a) $\Psi \circ \lambda(x) \circ \Psi^{-1} = \lambda(x) \otimes \text{id}$, $\Psi \circ \rho(x) \circ \Psi^{-1} = \rho(x) \otimes \text{Ad } x$ $(x \in G)$;

(b) (*notations of* 3.3.2(b)) $\Psi^{-1}(g \otimes X)(f) = - \sum_{i=1}^{r} (gf_i).(Xg_i)$. This is a consequence of 3.3.2 and 3.2.2(i). These results yield an isomorphism $\phi \mapsto \phi \circ \Phi \circ d$ of Ω_G onto $\mathrm{Hom}_{k[G]}(k[G] \otimes_k (T_eG)^*, k[G])$, and the latter module is isomorphic to $k[G] \otimes_k T_eG$ (the isomorphism sending $f \otimes X \in k[G] \otimes T_eG$ to the homomorphism ϕ: $k[G] \otimes (T_eG)^* \to k[G]$ with $\phi(g \otimes u) = u(X)fg$, $f,g \in k[G], X \in T_eG$, $u \in (T_eG)^*$). The assertions of 3.3.4 are then readily checked. For the last one observe that $Xg_i = \delta g_i(X)$ (notations of 3.3.2).

Let α_G: $\mathcal{D}_G \to T_eG$ be the linear map with $(\alpha_G D)f = (Df)e$.

3.3.5. *Proposition*. (i) α_G *defines an isomorphism of vector spaces* β_G: $L(G) \xrightarrow{\sim} T_eG$. *We have* $\beta_G \circ \rho(x) \circ \beta_G^{-1} = \mathrm{Ad}\ x$;
(ii) Ad *is a rational representation of* G *in* T_eG (*the adjoint representation*).

Let Ψ be as in 3.3.4. It is immediate from 3.3.4 that $\Psi L(G) = 1 \otimes T_eG$. Moreover, it follows from 3.3.4(b) that (with the previous notations)

$$(\alpha_G \circ \Psi^{-1})(1 \otimes X)f = - \sum_{i=1}^{r} f_i(e)\ Xg_i = -Xf,$$

since $f = \sum_i f_i(e) g_i$. Now (i) and (ii) readily follow.

3.3.6. Next let G' be a closed subgroup of G. Let $I \subset k[G]$ be the ideal of functions vanishing on G'. So we may assume that $k[G'] = k[G]/I$. Put

$$\mathcal{D}_{G,G'} = \{D \in \mathcal{D}_G | DI \subset I\}.$$

Then $\mathcal{D}_{G,G'}$ is a subalgebra of the Lie algebra $\mathcal{D}_G$ and there is

an obvious homomorphism of Lie algebras $\lambda: \mathcal{D}_{G,G'} \to \mathcal{D}_{G'}$. Also, $T_eG' = \{X \in T_eG | XI = 0\}$.
It follows from the definitions that

$$\alpha_G | \mathcal{D}_{G,G'} = \alpha_{G'} \circ \lambda.$$

3.3.7. <u>Lemma</u>. <u>λ defines an isomorphism of $\mathcal{D}_{G,G'} \cap L(G)$ onto $L(G')$.</u>

The injectivity of λ follows from the injectivity of β_G. To prove surjectivity we have to show that if $X \in T_eG'$ we have $\Psi^{-1}(1 \otimes X) \in \mathcal{D}_{G,G'}$ (Ψ as in 3.3.4). Let $f \in I$ and put, as before, $f(xy) = \sum_{i=0}^{r} f_i(x)g_i(y)$. Since $f(xy) = 0$ if $x,y \in G'$ and since the ideal defining $G' \times G'$ in $k[G \times G] = k[G] \otimes_k k[G]$ is $I \otimes k[G] + k[G] \otimes I$, we may assume that for each $i=1,\dots,r$ we have $f_i \in I$ or $g_i \in I$. By 3.3.4(b) we have

$$(\Psi^{-1}(1 \otimes X)f)(x) = - \sum_{i=1}^{r} f_i(x).(Xg_i).$$

If $x \in G'$ then all terms in the righthand side vanish. This proves the lemma.

Henceforth we identify the Lie algebra $L(G)$ and the tangent space T_eG, via Ψ. We thus obtain a Lie algebra structure on the latter space. We shall also denote the Lie algebra of a linear algebraic group $G,H,\dots$ by a corresponding gothic letter $\mathfrak{g},\mathfrak{h},\dots$.

If $\phi: G \to G'$ is a homomorphism of algebraic groups, we write $d\phi$ for the tangent map $(d\phi)_e: \mathfrak{g} \to \mathfrak{g}'$.

We call $d\phi$ the differential of ϕ.

3.3.8. Proposition. $d\phi$ is a homomorphism of Lie algebras, which is compatible with the p-operations if $p > 0$.

Using a factorization of ϕ

$$G \overset{\rho}{\to} G \times G' \overset{\sigma}{\to} G',$$

where $\rho = (x, \phi x)$ and σ is projection onto the second factor one sees that it suffices to prove the proposition in the three cases that ϕ is an isomorphism, a projection σ, or an injection of a closed subgroup. It is left to the reader to check the first two cases. In the last case the proposition follows from 3.3.6 and 3.3.7.

3.3.9. Examples.

(1) Let $G = \mathbb{G}_a$, so $k[G] = k[T]$. The derivations of $k[G]$ commuting with all translations $T \mapsto T+a$ ($a \in k$) are the k-multiples of $X = \frac{d}{dT}$. Clearly $X^p = 0$ (if $p > 0$).
So $\mathfrak{g}$ is the 1-dimensional Lie algebra kX, with $[X,X] = 0$ and $X^p = 0$ if $p > 0$.

(2) Let $G = \mathbb{G}_m$, then $k[G] = k[T,T^{-1}]$. The derivations of $k[G]$ commuting with the translations $T \mapsto aT$ ($a \in k^*$) are the k-multiples of $X = T\frac{d}{dT}$. We have $X^p = X$ (if $p > 0$).
Now $\mathfrak{g}$ is the 1-dimensional Lie algebra kX, with $[X,X] = 0$ and $X^p = X$ if $p > 0$.

(3) If $G = \mathbb{GL}_n$ then $k[G] = k[T_{ij}, D^{-1}]_{1 \leq i,j \leq n}$, where $D = \det(T_{ij})$ (see 2.1.3(3)). Denote by $\mathfrak{gl}_n$ the Lie algebra of all n×n-matrices over k, with product $[X,Y] = XY-YX$ (and X^p the usual p-power). If $X = (x_{ij}) \in \mathfrak{gl}_n$,

$$D_X T_{ij} = - \sum_{k=1}^{n} T_{ih} x_{hj}$$

defines a derivation of k[G] commuting with all $\lambda(x)$ $(x \in G)$. Since the map $X \mapsto D_X$ is injective, and since dim $\mathfrak{g}$ = dim G = n^2 = dim $\mathfrak{gl}_n$, we must have that L(G) consists of the D_X. So $L(\mathbb{GL}_n) \simeq \mathfrak{gl}_n$. Then $Ad(x)X = xXx^{-1}$ $(x \in \mathbb{GL}_n, X \in \mathfrak{gl}_n)$. Also, if G is a closed subgroup of $\mathbb{GL}_n$, we may view $\mathfrak{g}$ as a subalgebra of $\mathfrak{gl}_n$ (check these facts).

3.3.10. Exercises. (1) The Lie algebra of $\mathbb{SL}_n$ is the subalgebra $\mathfrak{sl}_n$ of $\mathfrak{gl}_n$ consisting of the matrices with trace 0 (Hint: use 3.3.7).

(2) Determine the Lie algebras of the groups $\mathbb{D}_n, \mathbb{T}_n, U_n$ of 2.1.3.

(3) Let $\phi: \mathbb{SL}_2 \to \mathbb{PSL}_2$ be the homomorphism of 2.1.4. Show that $d\phi$ is surjective if and only if $p \neq 2$.

(4) If $p \neq 2$ then $L(\mathbb{SO}_n)$ is isomorphic to the Lie algebra of all skew-symmetric n×n-matrices.

(5) $L(G) = L(G^0)$.

(6) Show that Ad(x) is an automorphism of the Lie algebra L(G) $(x \in G)$.

If $G = \mathbb{GL}_n$ then $Ad(x)X = xXx^{-1}$ $(x \in \mathbb{GL}_n, X \in \mathfrak{gl}_n)$.

We next give some differentiation formulas, to be used in the sequel.

Let G be a linear algebraic group, let π: G × G be the product morphism and $\iota: G \to G$ inversion (see 2.1.1). We have $L(G \times G) = \mathfrak{g} \oplus \mathfrak{g}$ (3.1.7(3)).

3.3.11. <u>Lemma</u>. $(d\pi)_{(e,e)}: \mathfrak{g} \oplus \mathfrak{g} \to \mathfrak{g}$ <u>is the map</u> $(X,Y) \to X+Y$ <u>and</u> $(d\iota)_e X = -X$.

π defines a k-linear map $\tilde{\pi} = (\pi^*)^0$ of Ω_G to $\Omega_{G\times G}$ (see 3.2.3) and $\Omega_{G\times G} = \Omega_G \otimes k[G] \oplus k[G] \otimes \Omega_G$ (3.2.5(6)).

Let $f \in k[G]$, $\pi^* f = \sum_i f_i \otimes g_i$. Then $\tilde{\pi}(df) = \sum_i (df_i \otimes g_i + f_i \otimes dg_i)$.

Since $\sum_i f_i(e) g_i = \sum_i f_i g_i(e) = f$, we have

$$\tilde{\pi}(df) - df \otimes 1 - 1 \otimes df \in M_{(e,e)} \Omega_{G\times G}.$$

It follows that the linear map induced by $\tilde{\pi}$ of $\Omega_G(e)$ to $\Omega_{G\times G}(e,e) = \Omega_G(e) \oplus \Omega_G(e)$ sends u to (u,u). Since the dual of this map is $(d\pi)_{(e,e)}$, the first assertion follows.

The last assertion follows by using that $\pi \circ (\mathrm{id},\iota)$ is the trivial map $G \to e$ (check this).

3.3.12. <u>Lemma</u>. (i) <u>Let</u> $\sigma: G \to G$ <u>be a morphism of algebraic varieties</u>, <u>put</u> $\phi x = (\sigma x)x^{-1}$. <u>Then</u> $(d\phi)_e = (d\sigma)_e - 1$;

(ii) <u>Let</u> $x \in G$, <u>put</u> $\phi_x y = xyx^{-1}y^{-1}$. <u>Then</u> $(d\phi_x)_e = \mathrm{Ad}(x) - 1$.

This follows from 3.3.11. The details can be left to the reader.

If V is a finite dimensional vector space we write $\mathfrak{gl}(V)$ for the Lie algebra of all endomorphisms of V. So $\mathfrak{gl}(V) \simeq \mathfrak{gl}_n$, if $n = \dim V$. If $\phi: G \to GL(V)$ is a rational representation (2.3.1) then its differential $d\phi$ is a homomorphism of Lie algebras $\mathfrak{g} \to \mathfrak{gl}(V)$, i.e. a representation of $\mathfrak{g}$ in V.

Now let G_1 and G_2 be two linear algebraic groups and let ϕ_i be a rational representation of G_i in V_i (i = 1,2). Let

$\phi_1 \oplus \phi_2$ and $\phi_1 \otimes \phi_2$ be the direct sum and tensor product representations of $G_1 \times G_2$ in $V_1 \oplus V_2$ and $V_1 \otimes V_2$. We identify $L(G_1 \times G_2)$ with $\mathfrak{g}_1 \oplus \mathfrak{g}_2$.

3.3.13. Lemma. (i) $d(\phi_1 \oplus \phi_2) = d\phi_1 \oplus d\phi_2$;
(ii) $d(\phi_1 \otimes \phi_2)(X_1,X_2)(v_1 \otimes v_2) =$
$(d\phi_1)(X_1)v_1 \otimes v_2 + v_1 \otimes (d\phi_2)(X_2)v_2$.
We only prove (ii), the proof of (i) is similar. It suffices to prove (ii) for $X_2 = 0$. In that case we have that $d(\phi_1 \otimes \phi_2)(X_1,0)(v_1 \otimes v_2)$ is the differential of the morphism $x \mapsto \phi_1(x)v_1 \otimes v_2$ of G_1 to $V_1 \otimes V_2$, evaluated at X_1. The required formula then easily follows.

3.3.14. Exercises. (1) Let $\phi: G \to \mathbb{GL}_n$ be a rational representation, put $\phi(x) = (f_{ij}(x))$, with $f_{ij} \in k[G]$. Then $d\phi(X) = (Xf_{ij})$ $(X \in T_eG)$.
(2) Let $V \subset k[G]$ be a finite dimensional subspace of $k[G]$ which is stable under all $\lambda(x)$, $x \in G$ (all $\rho(x)$). Let $\phi: G \to GL(V)$ be the rational representation defined by λ (ρ, respectively). Then $(d\phi(X))f = Xf$ (where $X \in \mathfrak{g}$ is viewed as an element of $L(G) \subset \mathcal{D}_G$).
(3) The differential of the adjoint representation is given by $(d\,\mathrm{Ad})(X)Y = [X,Y]$ $(X,Y \in \mathfrak{g})$ (Hint: first deal with the case $G = \mathbb{GL}_n$ and then use an imbedding $G \hookrightarrow \mathbb{GL}_n$).
(4) Let $\phi: G \to GL(V)$ be a rational representation. Let $\Lambda^h V$ be an exterior power of V (see [26 , Ch. XVI, §6]), and let $\Lambda^h\phi: G \to GL(\Lambda^h V)$ be the representation with

$$(\Lambda^h\phi)(x)(v_1 \wedge \dots \wedge v_h) = \phi(x)v_1 \wedge \dots \wedge \phi(x)v_h \quad (x \in G, v_i \in V).$$

Then $\Lambda^h\phi$ is a rational representation and

$$(d\Lambda^h\phi)(X)(v_1 \wedge \ldots \wedge v_h) = \sum_{i=1}^{h} v_1 \wedge \ldots \wedge (d\phi)(X)v_i \wedge \ldots \wedge v_h$$

$$(x \in \mathfrak{g}, v_i \in V).$$

(Hint: deduce this from 3.3.13.(ii)).

3.3.15. As an application of 3.3.12(i) we shall prove Lang's theorem about algebraic groups over finite fields. To prepare for it, we must say words about fields of definition of varieties, which we have avoided so far.
Let X be an affine variety over k. Let k_0 be a subfield of k. A k_0-structure on X is a k_0-subalgebra A_0 of $A = k[X]$, of finite type over k_0, such that $A = A_0 \otimes_{k_0} k$. If such a structure is given we say that X is a k_0-variety, or is defined over k_0, or that k_0 is a field of definition of X.
A closed subvariety of $\mathbb{A}^n$ is defined over k_0 if and only if the ideal $I(X) \subset k[T_1,\ldots,T_n]$ (see 1.1.1) is generated by polynomials which are contained in $k_0[T_1,\ldots,T_n]$ (check this). This is not equivalent to the weaker property that there are $f_1,\ldots,f_s \in k_0[T_1,\ldots,T_n]$ such that X is the set of common zeros of the f_i $(1 \leqslant i \leqslant s)$ (counterexample: char $k = p > 0$, k_0 is a non-perfect field, $n = s = 1$ and $f_1 = T^p - a$, where $a \in k_0$, $a^{\frac{1}{p}} \notin k_0$). However, this equivalence is true if k_0 is perfect. For a discussion of these questions see e.g. [3 , AG, §12]. We do not need to go into them.

If X and Y are affine k_0-varieties, defined by k_0-algebras A_0 and B_0, a morphism $\phi: X \to Y$ is a k_0-morphism, or is defined

<u>over</u> k_0 if there is a homomorphism of k_0-algebras $\phi_0^*: B_0 \to A_0$, such that the homomorphism $\phi^*: k[Y] \to k[X]$ defining ϕ is $\phi_0^* \otimes \mathrm{id}$. It is now clear how to define a k_0-structure on a linear algebraic group G. In that case we say that G is a k_0-group or is defined over k_0.

Now let $k_0 = \mathbb{F}_q$, the finite field with q elements, and let k be an algebraic closure of $\mathbb{F}_q$. If X is an affine variety defined over k_0, and A, A_0 are as before, the homomorphism $f \mapsto f^q$ of A_0 into A_0 defines a morphism $F: X \to X$ which is defined over $\mathbb{F}_q$. This is the <u>Frobenius morphism</u>.

If $\phi: X \to \mathbb{A}^n$ is an $\mathbb{F}_q$-imbedding of X in affine n-space, so that we can view X is a subset of k^n, we have $F(x_1,\dots,x_n) = (x_1^q,\dots,x_n^q)$, if $(x_1,\dots,x_n) \in X$.

It is clear from the definition of F that all tangent maps $(dF)_x$ are zero.

We can now prove Lang's theorem.

3.3.16. <u>Theorem</u>. <u>Let</u> G <u>be a connected linear algebraic group which is defined over</u> $\mathbb{F}_q$, <u>let</u> F <u>be the corresponding Frobenius morphism</u>. <u>Then</u> $\phi: x \mapsto (Fx)x^{-1}$ <u>is a surjective morphism of</u> G.

As a consequence of 3.3.12(i) we see that $(d\phi)_e$ maps T_eG bijectively onto T_eG. Using that $\phi x = (Fa)\phi(a^{-1}x)a^{-1}$ it then follows that $(d\phi)_a$ maps T_aG bijectively onto $T_{\phi a}G$, for all $a \in G$.

Let X be the closure $\overline{\phi G}$, this is an irreducible closed subvariety of G. It follows from 1.9.5 and 3.2.17(ii) that there exists $a \in G$ such that ϕa is a simple point of X. The preceding remarks then imply that $\dim T_{\phi a}X = \dim G$. Since $\dim T_aX = \dim X$, we have $X = G$ (1.8.2). Hence (1.9.5) ϕG contains a non-empty open subset U of G.

Let $a \in G$. A similar argument shows that there is a non-empty open subset V of G consisting of elements of the form $(Fy)ay^{-1}$. Since G is connected we have $U \cap V \neq \phi$, and there exist $x,y \in G$ such that $(Fx)x^{-1} = (Fy)ay^{-1}$. This implies the theorem.

If G is as in the theorem, the fixed point set of F

$$G^F = \{x \in G | Fx = x\}$$

is a finite group (the "finite Chevalley groups" are all of this kind). Lang's theorem is a basic tool in the study of these groups. We give an instance of an application in the next exercise.

3.3.17. Exercises. (1) Let G and G^F be as before. Let $x \in G^F$, then its centralizer $Z_G(x)$ is a closed subgroup of G (see 2.5), which is F-stable. Assume that $Z_G(x)$ is connected, if $y \in G^F$ is conjugate to x in G then it is already conjugate to x in the finite group G^F (for more details see [6 , part E, Ch. I]).

(2) Let G be a connected linear algebraic group and $\sigma: G \to G$ an endomorphism of algebraic groups such that $d\sigma: \mathfrak{g} \to \mathfrak{g}$ is a nilpotent linear map. Then $x \mapsto (\sigma x)x^{-1}$ is surjective.

3.4. Jordan decomposition in $\mathfrak{g}$.

Let G be a linear algebraic group, with Lie algebra $\mathfrak{g}$. We

view the elements of $\mathfrak{g}$ as derivations of k[G].
There is a Jordan decomposition of elements of $\mathfrak{g}$, quite similar to the one in G 2.4.8 , which will be briefly discussed.

3.4.1. Lemma. *Let F be a finite dimensional subspace of* k[G]. *There is a finite dimensional subspace* E *of* k[G] *which contains* F *and is stable under all derivations* $X \in \mathfrak{g}$.

From this it follows that we have an additive Jordan decomposition $X = X_s + X_n$, where X_s and X_n are linear maps of k[G], which are semi-simple and locally nilpotent, respectively (2.4.6).

3.4.2. Theorem. (i) X_s *and* X_n *lie in* $\mathfrak{g}$ *and* $[X_s, X_n] = 0$;
(ii) *If* $\phi: G \to G'$ *is a homomorphism of linear algebraic groups then* $((d\phi)X)_s = (d\phi)X_s, ((d\phi)X)_n = (d\phi)X_n$;
(iii) *If* $G = \mathbf{GL}_n$ *then* X_s *and* X_n *are the semi-simple and nilpotent parts of the matrix* $X \in \mathfrak{gl}_n$.

3.4.3. Exercises. (1) Prove 3.4.1 and 3.4.2 (Hint: the proofs are similar to those of 2.3.4 and 2.4.8, respectively).
(2) If G is a torus then all $X \in \mathfrak{g}$ are semi-simple.
(3) If G is unipotent then all $X \in \mathfrak{g}$ are nilpotent.

Notes.

3.1 and 3.2 contain standard material on tangent spaces and simple points of algebraic varieties. The basic geometric results are 3.2.17 and 3.2.20. Proofs of the relevant algebraic results about separability of field extensions have also been included.

Our treatment makes much use of modules of differentials. We have tried to keep the discussion of their formal properties as brief as possible. A thorough discussion can be found in [20 , Ch.0,§20]. The results of 3.2.16 are taken from [20 , Ch.0, 19.1.12].

In the discussion of the Lie algebra of a linear algebraic group G in 3.3 also much use is made of differentials. We have found it quite convenient to derive the basic properties of the Lie algebra of G (for example those of 3.3.5) from the formal properties of the differential module Ω_G, given in 3.3.2.

3.3.15 is the only place in these notes where non algebraically closed groundfields occur. We have to introduce them there to prepare for Lang's theorem 3.3.16.

4. Further study of morphisms, applications.

4.1. General properties of morphisms of varieties.

4.1.1. We shall study the following situation. X and Y are two irreducible varieties and $\phi: X \to Y$ is a dominant morphism (see 3.2.19). We may view the quotient field k(X) (see 1.8.1) as an extension of the field k(Y). The transcendence degree $\mathrm{trdeg}_{k(Y)}k(X)$ equals dim X - dim Y. We shall give in 4.1.6(ii) a geometric interpretation of this integer.
If F is any field and E a finite algebraic extension, the elements of E which are separable over F form a subfield E_s, which is a separable extension of F. Its degree $[E_s:F]$ is the <u>separable degree</u> of the extension E/F, also denoted $[E:F]_s$. For any $x \in E$, there is a power $x^{p^e} \in E_s$, where p = char F (see [26 , Ch. VII, §4]).
If X and Y are as before and dim X = dim Y, then k(X) is a finite algebraic extension of k(Y). In 4.1.6(iii) a geometric interpretation is given of $[k(X):k(Y)]_s$.
If k(X) = k(Y) then ϕ is said to be <u>birational</u>.

4.1.2. <u>Lemma</u>. *ϕ is birational if and only if there is a non-empty open subset U of X such that ϕU is open in Y and that ϕ induces an isomorphism of varieties $U \xrightarrow{\sim} \phi U$.*
It follows from the definition of quotient fields (1.8.1) that ϕ is birational if the condition is satisfied. So assume ϕ to be birational. We may then take X and Y to be affine (check this). Let $k[X] = k[Y][f_1,\dots,f_r]$, where $f_i \in k(Y)$. Take $f \in k[Y]$ such that $f \neq 0$, $ff_i \in k[Y]$. It is now clear that ϕ

induces an isomorphism $k[Y]_f \tilde{\to} k[X]_f$, whence an isomorphism of varieties $D_X(f) \tilde{\to} D_Y(f)$ (see 1.4.6).

The main result of this section is 4.1.6. We first discuss some particular cases. Assume that X and Y are affine, and that there is $a \in k[X]$ such that $k[X] = k[Y][a]$.

4.1.3. Lemma. *Let a be transcendental over k(Y).*

(i) *ϕ is an open morphism;*

(ii) *If Y' is an irreducible closed subvariety of Y, then $\phi^{-1}Y'$ is an irreducible closed subvariety of X, of dimension* dim Y'+1.

Recall that an open map ϕ: X → Y of topological spaces is one which maps open sets onto open sets.

We may assume that $X = Y \times \mathbb{A}^1$ and that ϕ is projection onto the first factor. Let $f = \sum_{i=0}^{r} f_i T^i \in k[X] = k[Y][T]$. Then

$$\phi(D_X(f)) = \bigcup_{i=0}^{r} D_Y(f_i),$$

whence (i).

If Q is the prime ideal of the irreducible closed subvariety Y' of Y, then $\phi^{-1}Y'$ is the set of points of k[X] in which the functions of the ideal $P = Qk[X] = \{\sum_{i\geq 0} f_i T^i | f_i \in Q\}$ all vanish. It is clear that $k[X]/P \simeq (k[Y]/Q)[T]$. Since the last ring is an integral domain P is a prime ideal. Hence $\phi^{-1}Y'$ is irreducible. The last point of (ii) also follows readily.

4.1.4. Lemma. *Let a be separably algebraic over k(Y). There is a non-empty open subset U of X with the following properties:*

(i) *The restriction of ϕ to U is an open morphism U → Y;*

(ii) _If_ Y' _is an irreducible closed subvariety of_ Y _and_ X' _is an irreducible component of_ $\phi^{-1}Y'$ _such that_ $X' \cap U \neq \phi$, _then_ $\dim X' = \dim Y'$;

(iii) _If_ $x \in U$, _the number of points of the fiber_ $\phi^{-1}(\phi x)$ _equals the degree_ $[k(X):k(Y)]$.

We have $k[X] = k[Y][T]/I$, where I is the ideal in $k[Y][T]$ of the polynomials F with $F(a) = 0$. Let f be the minimum polynomial of a over $k(Y)$ (i.e. the irreducible polynomial in $k(Y)[T]$ with leading coefficient 1 and of lowest degree, which has a as a root). Let $b \neq 0$ be an element of $k[Y]$ such that all coefficients of f lie in $k[Y]_b$.

Let $x_1,\dots,x_n$ be the roots of f, in some extension field of $k(Y)$. Since a is separable over $k(Y)$, the x_i are all distinct, and $c = \prod_{i<j} (x_i-x_j)^2$ (the "discriminant" of f, see e.g. [26 , Ch. V, §9]) is a nonzero element of $k[Y]_b$.

Replace X and Y by $D_X(bc)$ and $D_Y(bc)$, respectively. We then are reduced to proving the lemma when, moreover, the following holds:

(a) I contains the minimum polynomial f. From this it follows, using the division algorithm, that I is precisely the ideal generated by f. It also follows that $k[X]$ is a free $k[Y]$-module.

(b) If $f(T) = \sum_{i=0}^{n} f_i T^i$, then for all $y \in Y$ the polynomial $f(y)(T) = \sum_{i=0}^{n} f_i(y)T^i$ has distinct roots.

We shall show that in this situation the assertions of the lemma hold, with $U = X$.

We may assume that

$$X = \{(y,t) \in Y \times \mathbb{A}^1 | f(y)(t) = 0\},$$

φ being projection on Y.

Let $g \in k[Y][T]$ and denote by g' be its image in $k[X]$. Then

$$D_X(g') = \{(y,t) \in X | g(y)(t) \neq 0\}.$$

Write $g = gf+r$, where $\deg r < \deg f = n$, so $r = \sum_{i=0}^{n-1} r_i T^i$.
It follows that $\phi D_X(g')$ is the set of $y \in Y$ such that not all roots of $f(y)(T)$ are roots of $r(y)(T)$. Since the first polynomial has no multiple roots, this means that

$$\phi D_X(g') = \bigcup_{i=0}^{n-1} D_Y(r_i),$$

whence (i).

Next let Y' be as in (ii), and let Q be the corresponding prime ideal in $k[Y]$. Then $\phi^{-1}Y'$ is the closed set defined by the ideal $Qk[X]$. Putting $B = k[Y]/Q$, we have
$k[X]/Qk[X] \simeq B[T]/(\overline{f})$, where $\overline{f}$ is the image of f in $B[T]$.
We prove that $Qk[X]$ is a radical ideal, i.e. that $B[T]/(\overline{f})$ is a reduced algebra (see 1.3.1). Let $F \in B[T]$ and assume that F^h is divisible by $\overline{f}$ for some $h > 0$. We may assume that $\deg F < \deg \overline{f} = n$. But F^h is divisible by $\overline{f}$ in the polynomial algebra over the quotient field of B. Since, as a consequence of (b), $\overline{f}$ has distinct roots, we must have that F is divisible by $\overline{f}$. Since $\deg F < \deg \overline{f}$ it follows that $F = 0$. So $Qk[X]$ is a radical ideal. By 1.2.6(2) it is an intersection of prime ideals, say $Qk[X] = P_1 \cap \ldots \cap P_r$. We may assume that there are no inclusions among the P_i. The irreducible components of $\phi^{-1}Y'$ are the sets $V_X(P_i)$. We claim that $P_i \cap k[Y] = Q$ $(1 \leq i \leq r)$.

If this is not so, let, e.g., $P_1 \cap k[Y] \neq Q$. Take $x_1 \in P_1 \cap k[Y] - Q$, and x_i in $P_i - P_1$ $(2 \leq i \leq r)$. Then $x_1 x_2 \ldots x_r \in P_1 \cap \ldots \cap P_r = Qk[X]$. Since $k[X]$ is free over $k[Y]$ it follows that we must have $x_2 \ldots x_r \in Qk[X] \subset P_1$, which is impossible. This establishes our claim.

It follows that the quotient field of $k[X]/P_i$ is an algebraic extension of B, proving (ii).

If Y' is reduced to a point, Q is a maximal ideal of $k[Y]$, and $B = k$. The preceding analysis shows that $\phi^{-1}Y'$ is the 0-dimensional variety defined by the reduced k-algebra $k[T]/(\overline{f})$, which implies (iii).

4.1.5. Lemma. *Let $p = \operatorname{char} k > 0$ and assume that $a^p \in k(Y)$. There is a non-empty open subset U of X with the following properties:*

(i) The restriction of ϕ to U is an open morphism $U \to Y$ which defines a homeomorphism $U \simeq \phi U$;

(ii) If Y' is an irreducible closed subset of Y there is at most one irreducible component X' of $\phi^{-1}Y'$ such that $X' \cap U \neq \phi$. If X' exists we have $\dim X' = \dim Y'$.

Let $a^p = b$. Replacing X and Y by $D_X(c)$ and $D_Y(c)$, with suitable $c \in k[Y]$, we are reduced to the case that $b \in k[Y]$ and that $k[X]$ is free over $k[Y]$. We shall prove in that case (i) and (ii) for $U = X$. The proof is similar to that of 4.1.4.

We can view X as the set of points $(y, b(y)^{\frac{1}{p}}) \in Y \times \mathbb{A}^1$ and ϕ as projection on Y. As in the proof of 4.1.4, one shows that ϕ is open. Since ϕ is clearly bijective, (i) follows.

Let Y' be as in (ii) and denote by Q the corresponding prime ideal in k[Y]. Put $B = k[Y]/Q$ and let $\bar{b} \in B$ be the image of b. We have $k[X]/Qk[X] \simeq B[T]/(T^p-\bar{b})$. If $T^p-\bar{b}$ is irreducible over the quotient field F of B, the latter ring is an integral domain, Qk[X] is a prime ideal and (ii) follows.
If $T^p-\bar{b}$ is reducible over F, it is of the form $(T-c)^p$. Then in $B[T]/(T^p-\bar{b})$ the elements d with $d^p = 0$ form a prime ideal (check this). It follows that $P = \{f \in k[X] \mid f^p \in Qk[X]\}$ is a prime ideal, and is the radical of Qk[X]. If $f \in P \cap k[Y]$ then $f^p \in Qk[X] \cap k[Y] = Q$ (since k[X] is free over k[Y]), whence $f \in Q$. So $P \cap k[Y] = Q$ and the assertions of (ii) follow, as in the proof of 4.1.4.

The next theorem is the main result of this section.

4.1.6. _Theorem_. _Let_ X _and_ Y _be irreducible varieties, let_ $\phi: X \to Y$ _be a dominant morphism_. _Put_ $r = \dim X - \dim Y$. _There is a non-empty open set_ U _in_ X _with the following properties_:

(i) _The restriction of_ ϕ _to_ U _is an open morphism_ $U \to Y$;

(ii) _If_ Y' _is an irreducible closed subvariety of_ Y _and_ X' _an irreducible component of_ $\phi^{-1}Y'$ _such that_ $X' \cap U \neq \emptyset$ _then_ $\dim X' = \dim Y' + r$. _In particular, if_ $y \in Y$, _any irreducible component of_ $\phi^{-1}y$ _which intersects_ U _has dimension_ r;

(iii) _If_ k(X) _is algebraic over_ k(Y) _then for all_ $x \in U$ _the number of points in the fiber_ $\phi^{-1}(\phi x)$ _equals_ $[k(X):k(Y)]_s$.

Assume we have a factorization $\phi = \phi' \circ \psi$, where $\psi: X \to X'$, $\phi': X' \to Y$ are dominant morphisms, X' being irreducible. It is easy to see that if (i) and (ii) hold for ϕ' and ψ, they also hold for ϕ (check this).

To prove the theorem we may assume X and Y affine. Since k[X] is a k[Y]-algebra of finite type, we can find a factorization of ϕ

$$X = X_r \xrightarrow{\phi_r} X_{r-1} \xrightarrow{\phi_{r-1}} \dots \longrightarrow X_1 \xrightarrow{\phi_1} X_0 = Y.$$

where the $k[X_i]$ are subalgebras of k[X] containing k[Y], and where the ϕ_i are as in the situations covered by one of the 3 preceding lemmas. (i) and (ii) then follow from these lemmas. A similar argument establishes (iii): first factorize ϕ as follows $X \overset{\psi}{\to} X' \overset{\phi'}{\to} Y$, where $k(X') = k(X)_s$ (if $k[X] = k[Y][a_1,\dots,a_s]$ one may take for X' the variety with coordinate algebra $k[Y][a_1^{p^{e_1}},\dots,a_s^{p^{e_s}}]$, the p-powers being chosen such that $a_i^{p^{e_i}}$ is separable over k(Y)).

The proof has the following useful corollary.

4.1.7. <u>Corollary</u>. <u>In the situation of</u> 4.1.6 <u>we may replace</u> (i) <u>by the stronger property</u>

(i)' <u>For any variety</u> Z, <u>the restriction of</u> ϕ <u>to</u> U <u>defines an open morphism</u> $U \times Z \to Y \times Z$.

It suffices to prove this for Z affine. Observe also that if (i)' holds for Z and if Z' is a closed subvariety of Z, it also holds for Z'. These remarks show that it suffices to establish (i)' for $Z = \mathbb{A}^n$, and this will follow if we prove (i)' with $Z = \mathbb{A}^n$ in the situations of 4.1.3, 4.1.4, 4.1.5. The latter is obvious in the first case. In the others, the following observation will imply (i)': the minimum polynomial of a over $k(Y)(T_1,\ldots,T_n)$ is the same as the minimum polynomial over $k(Y)$ (check this).

The content of theorem 4.1.6 is that a morphism ϕ as in 4.1.1 behaves "locally" in a nice way. That it need not do so globally can be seen from some of the next exercises. For more comments on the results of this section see the notes at the end of the chapter.
In the case of equivariant morphisms of homogeneous spaces, 4.1.6 suffices to get global good behaviour, see 4.3.3.

4.1.8. Exercises. (1) Show, using 4.1.6(iii), that an isomorphism $\phi\colon \mathbb{A}^1 \to \mathbb{A}^1$ is of the form $\phi t = at+b$ $(a,b \in k,\ a \neq 0)$. Deduce that an isomorphism ψ of the projective line $\mathbb{P}^1$ (see 1.7.1) is of the form $\psi(x_0,x_1)^* = (ax_0+bx_1,cx_0+dx_1)^*$, with $\begin{pmatrix} a & b \\ c & d \end{pmatrix} \in \mathbb{GL}_2$.
(2) Let $X = \{(x_1,x_2) \in \mathbb{A}^2 \mid x_1^3 = x_2^2\}$, this is an irreducible closed subset of dimension 1. Define $\phi\colon X \to \mathbb{A}^1$ by $\phi(x_1,x_2) = x_1^{-1}x_2$ if $(x_1,x_2) \neq (0,0)$, $\phi(0,0) = 0$. Show that ϕ is a morphism of varieties which is birational and bijective, but is not an isomorphism of varieties.
(3) Consider the morphism $\phi\colon \mathbb{A}^2 \to \mathbb{A}^2$ with $\phi(x_1,x_2)=(x_1,x_1x_2)$. Show that ϕ is birational, and that it is not an open morphism. Determine the components of the sets $\phi^{-1}x$, $x \in \mathbb{A}^2$.
(4) Define $\phi\colon \mathbb{A}^3 \to \mathbb{A}^3$ by $\phi(x_1,x_2,x_3) = (x_1,x_1x_2,x_3)$. Let $X = \{(x_1,x_2,x_3) \in \mathbb{A}^3 \mid x_2^2 = 1+x_1\}$. Show that X and $Y = \phi X$ are irreducible closed subvarieties of $\mathbb{A}^3$ of dimension 2.

Let

$$Y' = \{(x_1,x_2,x_3) \in Y | x_2 = x_1x_3,\ x_3^2 = 1+x_1\}.$$

Show that Y' is irreducible, closed, of dimension 1 and that, if char $k \neq 2$, $\phi^{-1}Y' \cap X$ has a component of dimension 0.

4.2. Finite morphisms.

4.2.1. Let B be a ring and A a subring. We say that B is *finite* over A if B is an A-module of finite type. We say that $b \in B$ is integral over A if it satisfies an equation

$$b^n + a_1b^{n-1} + \ldots + a_n = 0,\ a_i \in A.$$

If every $b \in B$ is integral over A then B is said to be *integral* over A.

4.2.2. *Lemma. Let B be an A-algebra of finite type. Then B is integral over A if and only if B is finite over A.*

Let B be finite over A, so $B = Ab_1 + \ldots + Ab_m$. Let $b \in B$. There exist $a_{ij} \in A$ $(1 \leq i,j \leq m)$ such that

$$bb_i = \sum_{j=1}^{m} a_{ij}b_j,\quad 1 \leq i \leq m.$$

So

$$\sum_{j=1}^{m} (\delta_{ij}b - a_{ij})b_j = 0.$$

It follows that $\det(\delta_{ij}b - a_{ij})b_j = 0$ for all j, whence $\det(\delta_{ij}b - a_{ij}) = 0$. This implies that b is integral over A. That B is finite over A if it is an A-algebra of finite type and is integral over A is readily verified.

It follows from 4.2.2 that, in general, the $b \in B$ which are integral over A form a subring. It also follows that if B is integral over A and C is integral over B, then C is integral over A.

Now let $\phi: X \to Y$ be a dominant morphism of affine irreducible varieties. We say that ϕ is _finite_ if $k[X]$ is finite over $k[Y]$.

4.2.3. _Proposition. A finite morphism ϕ is surjective and closed._

Let $B = k[X]$, $A = k[Y]$. We have $B = A[b_1,\ldots,b_h]$, and an easy argument (see the proof of 4.1.6) shows that we may assume $h = 1$. But then 1.9.3 shows that every homomorphism of k-algebras $A \to k$ can be extended to a homomorphism $B \to k$. This means that ϕ is surjective. Let X' be an irreducible closed subset of X, whose prime ideal is Q. Then $\overline{\phi X'}$ is the closed set in Y defined by the prime ideal $Q \cap A$. Since B/Q is integral over $A/Q \cap A$, it follows from the surjectivity of $\phi|X'$ that $\phi X'$ is closed. This implies 4.2.3.

4.2.4. _Proposition. Let $\phi: X \to Y$ be a dominant morphism of irreducible affine varieties. Let $x \in X$ be such that its fiber $\phi^{-1}(\phi x)$ is finite. There is an affine open neighborhood U of ϕx in Y such that $\phi^{-1}U$ is an affine open neighborhood of x and that the restriction morphism $\phi^{-1}U \to U$ is finite._

If we have a factorization of $\phi: X \overset{\psi}{\to} X' \overset{\phi'}{\to} Y$ then clearly $\psi^{-1}(\psi x)$ is finite. If the assertion of the proposition is true for ψ, then replacing X and X' by suitable affine open sets, we may assume ψ to be surjective, by 4.2.3. But then we also have that $(\phi')^{-1}(\phi'\psi x)$ is finite. From this observation we conclude that it is sufficient to consider the case that $k[X] = k[Y][b] \simeq k[Y][T]/I$, where I is the ideal of the $f \in k[Y][T]$ such that $f(b) = 0$.

Let $\varepsilon: k[Y] \to k$ be the homomorphism defining ϕx. Extend ε to a homomorphism $k[Y][T] \to k[T]$ sending T to T. If $\varepsilon I = 0$ then $\phi^{-1}(\phi x)$ would be infinite and if $\varepsilon I = k[T]$ we would have $\phi^{-1}(\phi x) = \emptyset$ (check these facts).

This shows that there is $f \in I$ of the form

$$f_n T^n + \dots + f_m T^m + \dots + f_0,$$

where $\varepsilon f_n = \dots = \varepsilon f_{m+1} = 0$, $\varepsilon f_m \neq 0$ and $m > 0$. Put $s = f_n b^{n-m} + \dots + f_m$, then $s \neq 0$. Since

$$(*) \quad s b^m + f_{m-1} b^{m-1} + \dots + f_0 = 0,$$

we see that b is integral over the subring $k[Y][s^{-1}]$ of $k(X)$. But since $s \in k[Y][b]$, it follows that s is integral over $k[Y][s^{-1}]$, i.e. that s is integral over $k[Y]$. Hence $k[Y]\, s$ is finite over $k[Y]$.

Factorize $\phi: X \overset{\psi}{\to} X' \overset{\phi'}{\to} Y$, where $k[X'] = k[Y][s]$, so ϕ' is finite. From (*) we see that b is integral over $k[X'][s^{-1}]$. It follows that the restriction of ψ defines a finite morphism $D_X(s) \to D_{X'}(s)$. Let U be an affine open set such that $(\phi')^{-1}U \subset D_{X'}(s)$. Then U has the required property.

4.2.5. _Corollary_. _In the situation of_ 4.2.4 _we have_ $\dim X = \dim Y$.

4.2.6. _Exercises_. (1) Let ϕ be as in 4.2.3. For any variety Z, the induced morphism $\phi \times \mathrm{id}: X \times Z \to Y \times Z$ is closed.
(2) Let $\phi: X \to Y$ be a surjective morphism of irreducible affine varieties. Assume that $\phi^{-1}(\phi x)$ is finite for all $x \in X$. Then ϕ is finite (Hint: Take $f \in k[X]$ and put $I = \{g \in k[Y] | gf$ is integral over $k[Y]\}$, show that $I = k[Y]$).
(3) An integral domain A is said to be _normal_ if every element of its quotient field which is integral over A, lies already in A. An irreducible affine variety X is normal, if $k[X]$ is normal.
Let $\phi: X \to Y$ be a morphism of irreducible affine varieties, which is bijective and birational. Assume Y to be normal. Show that ϕ is an isomorphism (This is a version of "Zariski's main theorem". Hint: use ex.(2)).

4.3. _Applications to algebraic groups_.

We shall now discuss the consequences of the results of 4.1 in the case of equivariant morphisms of homogeneous spaces. First we give some auxiliary results, which will also be needed later on. Let G be an algebraic group, let X be a G-space. See 2.3.1 for this notion, as well as for the notion of orbit.

4.3.1. _Lemma_. _Assume_ G _to be connected_.
(i) _An orbit of_ G _in_ X _is open in its closure. It is an irreducible variety_;

(ii) _There exist closed orbits, viz. all orbits of minimal dimension_.

Let $x \in X$ and put $O = G.x$. Applying 1.9.5 to the morphism $G \to X$ which sends g to $g.x$ we see that O contains a non-empty open subset U of its closure $\overline{O}$. Since G acts transitively on O, the translates $g.U$ cover O, and the first point of (i) follows. It also follows that O is an open subvariety of the variety $\overline{O}$. Since G is irreducible the same is true for O and $\overline{O}$ (1.2.3(i)), proving the second point of (i).

Take O such that $\dim O$ is minimal. If O were not closed then $\overline{O}-O$ contained an orbit with dimension less than $\dim \overline{O} = \dim O$, which is a contradiction. This establishes (ii).

Now let X be a homogeneous space for G. Denote, as usual, by G^0 the identity component of G.

4.3.2. _Lemma_. (i) _Each irreducible component of_ X _is a homogeneous space for_ G^0;

(ii) _The components of_ X _are open and closed and_ X _is their disjoint union_.

Let $X' \subset X$ be an orbit of G^0. Since G acts transitively on X, it follows from 2.2.1(i) that X is the disjoint union of finitely many translates $g.X'$. Each of them is a G^0-orbit. From 4.3.1(ii) it then follows that one of them is closed. Hence they are all closed. The assertions now follow readily.

4.3.3. _Theorem_. _Let_ G _be an algebraic group and_ $\phi: X \to Y$ _an equivariant morphism of homogeneous spaces for_ G. _Put_ $r = \dim X - \dim Y$.

(i) _For any variety_ Z, _the induced morphism_ $\phi \times id$: $X \times Z \to Y \times Z$ _is open_;

(ii) _If_ Y' _is an irreducible closed subvariety of_ Y _and_ X' _an irreducible component of_ $\phi^{-1}Y'$ _then_ dim X' = dim Y'+r. _In particular, if_ $y \in Y$ _all irreducible components of_ $\phi^{-1}y$ _have dimension_ r;

(iii) ϕ _is an isomorphism if and only if it is bijective and if for some_ $x \in X$ _the tangent map_ $(d\phi)_x: T_xX \to T_{\phi x}Y$ _is bijective_.

Using 4.3.2 one reduces the proof to the case that G is connected and X,Y are irreducible. Since ϕ is surjective, hence dominant, we can apply 4.1.6. Let $U \subset X$ be an open subset with the properties of 4.1.6 and 4.1.7. It is clear that all translates g.U also enjoy these properties and that the same is true for a union of translates. Since they cover X, we have (i) and (ii).

If ϕ is bijective we conclude from 4.1.6(iii) that the field extension k(X)/k(Y) is purely inseparable. If $(d\phi)_x$ is bijective, we see from 3.2.21(ii) that this extension also is separable. Hence k(X) = k(Y) and ϕ is birational. Using 4.1.2 and a covering argument as before we conclude that ϕ is an isomorphism of varieties, from which (iii) follows.

We record explicitly a special case.

4.3.4. _Corollary_. _Let_ $\phi: G \to G'$ _be a surjective homomorphism of algebraic groups_.

(i) dim G = dim G' + dim Ker ϕ;

(ii) ϕ _is an isomorphism if and only if it is an isomorphism of abstract groups and if the Lie algebra homomorphism_ $d\phi$: $L(G) \to L(G')$ _is an isomorphism_.

View G and G' as homogeneous spaces for G: the first one via left translations (see 2.3.1) and the second one by $g.g' = \phi(g)g'$. Then ϕ is an equivariant morphism, and we can apply the preceding theorem. The details are left to the reader.

4.3.5. Exercises. (1) The connectedness assumption in 4.3.1 can be omitted.

(2) (situation of 4.3.1) An orbit is a homogeneous space (this fact was already stated in 2.3.1).

(3) Let G be a unipotent linear algebraic group and X an affine G-space. Denote by O a G-orbit in X and let $\overline{O}$ be its closure. Assume that $\overline{O} \neq O$. By 4.3.1(i) and (1), $\overline{O}$-O is closed. Let I and I' be the ideals in k[X] of functions vanishing on $\overline{O}$ and $\overline{O}$-O, respectively. Then $I \subset I'$, $I \neq I'$ and $I' \neq k[X]$. Deduce from 2.4.11 that there exists $f \in I'-I$ such that $f(g.x) = f(x)$ $(g \in G, x \in X)$ and show that this leads to a contradiction.

Hence all orbits of G in X are closed ("theorem of Kostant-Rosenlicht").

(4) Give an example of a bijective homomorphism of algebraic groups $G \to G'$ which is not an isomorphism of algebraic groups.

(5) If char k = 0, a homomorphism of algebraic groups ϕ: $G \to G'$ is an isomorphism of algebraic groups if and only if it is an isomorphism of abstract groups. Similarly for homogeneous spaces.

(6) (a) Let $\sigma: G \to G'$ be a homomorphism of algebraic groups. Then $\dim(\mathrm{Ker}\,\sigma) + \dim(\mathrm{Im}\,\sigma) = \dim G$;
(b) Assume moreover that G is connected, that $G' = G$ and σ is surjective. Then $\mathrm{Ker}\,\sigma$ is a finite subgroup of the center of G (Hint for the proof of the last statement: let $g \in \mathrm{Ker}\,\sigma$ and consider the morphism $x \mapsto xgx^{-1}g^{-1}$ of G to $\mathrm{Ker}\,\sigma$).

4.4. Semi-simple automorphisms.

4.4.1. Let G be a connected linear algebraic group, let σ be an automorphism of G, in the sense of algebraic groups. We put $G_\sigma = \{x \in G \mid \sigma x = x\}$, this is a closed subgroup. Also, $\phi x = (\sigma x)x^{-1}$ defines a morphism of varieties $G \to G$.
The tangent map $d\sigma$ is an automorphism of the Lie algebra $\mathfrak{g}$ of G. We write $\mathfrak{g}_\sigma = \{X \in \mathfrak{g} \mid (d\sigma)X = X\}$.
Since $\phi G_\sigma = \{e\}$, we have $(d\phi)_e \mathrm{Lie}(G_\sigma) = \{0\}$. Using 3.3.12(i) we see that $\mathrm{Lie}(G_\sigma) \subset \mathfrak{g}_\sigma = \mathrm{Ker}(d\phi)_e$. Equality need not hold (see 4.4.9(1)). It will follow from 4.4.2 that equality does hold if char $k = 0$. Let G act on itself by $g.x = (\sigma g)xg^{-1}$ $(g,x \in G)$. Then ϕ induces a dominant morphism $\psi: G \to \overline{\phi G}$, which is G-equivariant (G acting on G by left translations and on $\overline{\phi G}$ in the way just defined). Also, ϕG is an orbit of G (for the action just defined) so, by 4.3.1(i), ϕG is open in $\overline{\phi G}$. It follows from 3.2.21(i) that e is a simple point of $\overline{\phi G}$. By 4.3.3(ii) we have $\dim \overline{\phi G} = \dim G - \dim G_\sigma$.

4.4.2. <u>Lemma</u>. *We have* $\mathrm{Lie}(G_\sigma) = \mathfrak{g}_\sigma$ *if and only if the morphism* $\psi: G \to \overline{\phi G}$ *is separable*.

Using 3.3.12(i) we obtain that

$$\dim \operatorname{Im}(d\psi)_e = \dim \mathfrak{g} - \dim \mathfrak{g}_\sigma \geqslant \dim \mathfrak{g} - \dim \operatorname{Lie}(G_\sigma) =$$
$$= \dim G - \dim G_\sigma = \dim \overline{\phi G}.$$

If $\operatorname{Lie}(G_\sigma) = \mathfrak{g}_\sigma$, we have equality everywhere and $(d\psi)_e$ is surjective, so that ψ is separable (3.2.21(ii)). Conversely, if ψ is separable then the outer members of the sequence of inequalites are equal, and we must have $\operatorname{Lie}(G_\sigma) = \mathfrak{g}_\sigma$.

We say that σ is <u>semi-simple</u> if the induced automorphism σ^* of $k[G]$ is semi-simple, in the sense of 2.4.7. Such automorphisms can be characterized in a more concrete way as follows.

4.4.3. <u>Lemma</u>. <u>σ is semi-simple if and only if there is an isomorphism τ of G onto a closed subgroup of some $\mathbb{GL}_n$ and a semi-simple element $s \in \mathbb{GL}_n$ normalizing τG such that</u>

$$\tau(\sigma x) = s\tau(x)s^{-1}, \ x \in G.$$

Let σ be semi-simple. Construct an isomorphism $\tau\colon G \to \mathbb{GL}_n$ as in the proof of 2.3.5, by using a suitable n-dimensional subspace E of $k[G]$. Using 2.3.4(i) it follows that we may assume that σ^* stabilizes E. Let $s \in \mathbb{GL}_n$ be the element defined by the restriction of σ^* to E. It is clear that τ and s are as required.

Conversely, if τ and s have the properties of the lemma it readily follows that σ is semi-simple, using that $k[G]$ is a quotient of $k[\mathbb{GL}_n]$ and that the inner automorphism of $\mathbb{GL}_n$ defined by s induces a semi-simple automorphism of $k[\mathbb{GL}_n]$ and of $k[G]$ (check the details).

The next theorem contains the main properties of semi-simple automorphisms.

4.4.4. <u>Theorem</u>. <u>Let</u> σ <u>be a semi-simple automorphism of the connected linear algebraic group</u> G, <u>put</u> $\phi x \doteq (\sigma x)x^{-1}$.

(i) ϕG <u>is closed and the induced morphism</u> $\psi: G \to \phi G$ <u>is separable</u>;

(ii) $\mathrm{Lie}(G_\sigma) = \mathfrak{g}_\sigma$.

By the previous lemma we may assume that G is a closed subgroup of $\mathbb{GL}_n$ and that there is a semi-simple $s \in \mathbb{GL}_n$ such that $\sigma x = sxs^{-1}$ $(x \in G)$. We first prove the separability of ψ. If $G = \mathbb{GL}_n$ then writing s as a diagonal matrix $\mathrm{diag}(a_1,\ldots,a_1,a_2,\ldots,a_2,\ldots,a_r,\ldots,a_r)$ (the a_i being the distinct eigenvalues of s) a direct computation which is left to the reader, shows that $\mathrm{Lie}(G_\sigma) = \mathfrak{g}_\sigma$, whence the separability of ψ by 4.4.2.

If G is arbitrary, extend σ to an automorphism of $\mathbb{GL}_n$, viz. the inner automorphism defined by s. Also, extend ϕ to a morphism $\mathbb{GL}_n \to \mathbb{GL}_n$. Let $X \in T_e\overline{\phi G}$. Since $T_e\overline{\phi G} \subset T_e\overline{\phi(\mathbb{GL}_n)}$ and since we have already dealt with the case of GL_n, it follows using 3.3.12(i) that there is $Y \in \mathfrak{gl}_n$ such that $X = (d\sigma-1)Y$. The semi-simplicity of σ implies that $d\sigma$ is a semi-simple automorphism of $\mathfrak{gl}_n$, stabilizing the subspace $\mathfrak{g}$ of $\mathfrak{gl}_n$. We can then write $\mathfrak{gl}_n$ as a direct sum of $d\sigma$-stable subspaces $\mathfrak{g} \oplus \mathfrak{g}'$, from which one concludes that we may take the above element Y to be in $\mathfrak{g}$. This means that $(d\psi)_e$ is surjective, and the separability of ψ follows. By 4.4.2 we know that (ii) holds, so it remains to be proved that ϕG is closed.

Put $m(T) = \prod_{i=1}^{s} (T-a_i)$. This is the minimum polynomial of s, the a_i being as above. Let $X \subset \mathbb{GL}_n$ be the set of elements x with the following properties: (a) x normalizes G, (b) $m(x) = 0$, (c) the characteristic polynomial of the restriction of Ad(x) to $\mathfrak{g}$ equals that of $d\sigma$.

? Then X is a closed subset containing s and s^{-1} and, since m has distinct roots, it follows that all $x \in X$ are semi-simple (check these facts).

If $x \in X$, let $G_x = \{g \in G | gxg^{-1} = x\}$ and $\mathfrak{g}_x = \{X \in \mathfrak{g} | \mathrm{Ad}(x)X = X\}$. By (ii), which we have already proved, it follows that $\dim G_x = \dim \mathfrak{g}_x$. But $\dim \mathfrak{g}_x$ equals the multiplicity of the eigenvalue 1 of the restriction of Ad(x) to $\mathfrak{g}$, which equals $\dim \mathfrak{g}_\sigma$ (by condition (c)). So $\dim G_x = \dim G_\sigma$ for all $x \in X$.

Now G operates on X by inner automorphisms, and 4.3.3(ii) (applied to the morphism $g \mapsto gxg^{-1}$, viewed as an equivariant morphism $G \to G$, for suitable actions of G on itself) shows that all orbits of G have dimension equal to $\dim G - \dim G_\sigma$. But then 4.3.1(ii) shows that all orbits must be closed. It then readily follows that ϕG is closed. This finishes the proof of 4.4.4.

4.4.5. _Corollary_. _Let_ G _be_ _a_ _connected_ _linear_ _algebraic_ _group_, _let_ $s \in G$ _be_ _semi-simple_.

(i) _The_ _conjugacy_ _class_ $C = \{xsx^{-1} | x \in G\}$ _is_ _closed_. _The_ _morphism_ $x \mapsto xsx^{-1}$ _of_ G _onto_ C _is_ _separable_;

(ii) _Let_ $Z = \{x \in G | xsx^{-1} = s\}$ _be_ _the_ _centralizer_ _of_ s _in_ G. _Then_ $\mathfrak{g} = (\mathrm{Ad}(s)-1)\mathfrak{g} \oplus \mathrm{Lie}(Z)$.

Let $\sigma x = s^{-1}xs$. By 4.4.3 σ is a semi-simple automorphism of G. If ϕ is as before, then $C = s(\phi G)$ and $Z = G_\sigma$. The corollary now follows from 4.4.4.

Conjugacy classes of arbitrary elements need not be closed. For an example see 4.4.9(6).

4.4.6. Assume that D is a diagonalizable linear algebraic group, which acts as a group of automorphisms on the connected group G. This means that G is a D-space in the sense of 2.3.1 and that for all $d \in D$ the morphism $g \mapsto d.g$ of G is an automorphism of G. From 2.5.2 we infer that these are semi-simple automorphisms.

Let

$$Z_G(D) = \{g \in G | d.g = g \text{ for all } d \in D\}.$$

D also acts as a group of automorphisms on the Lie algebra $\mathfrak{g}$, let

$$\mathfrak{z}_{\mathfrak{g}}(D) = \{X \in \mathfrak{g} | d.X = X \text{ for all } d \in D\}.$$

4.4.7. <u>Corollary</u>. Lie $Z_G(D) = \mathfrak{z}_{\mathfrak{g}}(D)$.

We use induction on dim G, starting with the case dim G = 0. If $Z_G(D) = G$ then D acts trivially on G and also on $\mathfrak{g}$, and the assertion is obvious. Now assume $Z_G(D) \neq G$ and choose $d \in D$ such that the dimension of the group $G_d = \{g \in G | d.g = g\}$ is as small as possible. Since D is commutative, we have that D stabilizes the identity component $H = G_d^0$, and that $Z_G(D) = Z_H(D)$. Applying 4.4.4(ii) and induction the assertion follows.

This corollary applies, in particular, if D is a diagonalizable subgroup of G which acts by inner automorphisms.

We finally give another application of 4.4.4(ii).

4.4.8. <u>Proposition</u>. <u>Let G be a connected nilpotent linear algebraic group. The set</u> G_s <u>of semi-simple elements is a central subgroup of</u> G.

If $x,y \in G$ write (x,y) for the commutator $xyx^{-1}y^{-1}$. Recall that G is nilpotent if there is $n > 0$ such that all repeated commutators $(x_1(\ldots(x_n,x_{n+1})\ldots)$ equal e.

Now let $s \in G$ be semi-simple and put $\sigma x = sxs^{-1}$, $\phi x = (s,x)$. If n is as before we have $\phi^n G = e$. Using 3.3.12(i) we see that $d\sigma - 1$ is a linear map of $\mathfrak{g}$ which is both semi-simple and nilpotent, hence it is 0. Then $d\sigma = 1$ and 4.4.4(ii) implies that $\sigma = \mathrm{id}$. Hence all semi-simple elements of G lie in the center of G and the assertion follows from 2.4.3(i).

A more precise result will be given later (6.8). It should be observed that the connectedness assumption in 4.4.8 is essential (see 4.4.9(5)).

4.4.9. <u>Exercises</u>. (1) Let char k = 2, let $G = \mathbb{SL}_2$. Denote by σ the inner automorphism of G defined by the element $\begin{pmatrix} 1 & 1 \\ 0 & 1 \end{pmatrix}$. Show that $\mathrm{Lie}(G_\sigma) \neq \mathfrak{g}_\sigma$.

(2) Let $G = \mathbb{GL}_n$, let σ be an inner automorphism of G. Show that $\mathrm{Lie}(G_\sigma) = \mathfrak{g}_\sigma$.

(3) Let G be a closed subgroup of $\mathbb{GL}_n$. The Lie algebra $\mathfrak{g}$ is a subalgebra of $\mathfrak{gl}_n$. Assume that there is a complementary

subspace $\mathfrak{g}'$ of $\mathfrak{g}$ in $\mathfrak{gl}_n$ which is stable under Ad G.

(a) If σ is an inner automorphism of G then Lie(G_σ) = $\mathfrak{g}_\sigma$ (Hint: imitate the argument of the second paragraph of the proof of 4.4.4).

(b) Let $x \in G$, let C be the conjugacy class of x in $\mathbb{GL}_n$. Show that $C \cap G$ consists of finitely many conjugacy classes of G (Hint: let X be a component of $C \cap G$, let $x \in X$ and denote by D the conjugacy class of x in G. Deduce from (a) that $T_xX = T_xD$ and that X = D).

(4) (a) Show that the number of unipotent conjugacy classes in $\mathbb{GL}_n$ is finite (use the theory of Jordan normal forms, see [26 , Ch. XV, §3]).

(b) Let char k $\neq$ 2. Let G = $\mathbb{SO}_n$, imbedded in $\mathbb{GL}_n$ in the usual manner. Then G satisfies the condition of the previous exercise.

Deduce from (a) and ex. 3 (b) that the number of unipotent conjugacy classes in G is finite (This conclusion can be drawn, by the same argument, in more general circumstances. See for example: R. Steinberg, Conjugacy classes in algebraic groups, Lect. Notes in Math. no. 366, Springer Verlag, 1974, p. 106. One gets involved here in the structure theory of reductive linear algebraic groups, discussed in our later chapters).

(5) Let G be the dihedral group of order 8. Then G is nilpotent and has an irreducible complex representation of degree 2. This shows that 4.4.8 cannot be extended to non-connected groups (similar properties hold for all non-abelian p-groups).

(6) (a) Show that the conjugacy class of $\begin{pmatrix} 1 & 1 \\ 0 & 1 \end{pmatrix}$ in $\mathbb{GL}_2$ is not

closed.

(b) More generally, one can show as follows that the conjugacy class of a unipotent element $x \neq 1$ in $\mathbb{GL}_n$ is not closed. Let $x = 1+X$, where X is a nonzero nilpotent n×n-matrix. Using the theory of Jordan normal forms, define a multiplicative 1-parameter subgroup λ of $\mathbb{GL}_n$ (2.5.1) such that $\lambda(x)X\lambda(x)^{-1} = x^2X$ ($x \in k^*$). Deduce from this that 1 lies in the closure of the conjugacy class of x.

Notes.

The main result 4.1.6 of 4.1 has been established by elementary methods. However, it is rather weak. For example, it does not imply the following basic result, in the situation of 4.1.6: $\dim \phi^{-1}y \geqslant r$ for all $y \in Y$. A consequence is established in 4.2.5, it is the only result of this kind needed later on (mainly in 7.3.1 and 8.1.4).

For a sharper result see [3 , p.38-39], see also [16 ,§4, no. 5]. A very thorough treatment of these matters can be found in [20, Ch. IV], see e.g. [loc. cit., 5.6, 13.1]. In such a treatment a considerable amount of commutative algebra is needed. Here we have tried to keep the commutative algebra as limited as possible. Therefore we have even avoided the use of the concept of normality (which is only introduced in ex. 4.2.6(3)).

There exists a connection between the "equidimensionality" property of 4.1.6(ii) and the "universal openness" property of 4.1.7. If Y is normal, these properties are equivalent (theorem of Chevalley, see [20 , Ch. IV, 14.4.1] for a precise formulation of a stronger result).

The example of 4.1.8(3), (4) are taken from [7 , ex.13,p.80].

The argument of the proof of 4.2.4 is due to Chevalley [10,p.177].

The result on semi-simple automorphisms of 4.4.4 is (seemingly) more general than the one which is usually given (see [3 , p.227-228]).

4.4.8 is deduced here directly from 4.4.4. The usual proof is somewhat different (see [3 , p.245-246].

5. Quotients.

In this chapter G will denote a linear algebraic group with Lie algebra $\mathfrak{g}$ and H a closed subgroup with Lie algebra $\mathfrak{h}$. We shall establish the existence of a quotient space G/H. Before doing this we prove some auxiliary results.

5.1. Chevalley's theorem.

5.1.1. Lemma. *There exists a finite dimensional $\rho(G)$-stable subspace V of k[G] together with a subspace W of V such that*

$$H = \{x \in G \mid \rho(x)W = W\},$$

$$\mathfrak{h} = \{X \in \mathfrak{g} \mid X.W \subset W\}.$$

Here $\rho(x)$ denotes right translation by x (see 2.3.1). Observe that if ϕ is the rational representation of G in V defined by ρ, we have for $f \in V$ that $d\phi(X).f = X.f$ (see 3.3.14(2)). Let $I \subset k[G]$ be the ideal of functions vanishing on H and take V to be a finite dimensional $\rho(G)$-stable subspace k[G] containing a set of generators $(f_1,\ldots,f_r)$ of I. Such a V exists by 2.3.4. Let $W = V \cap I$. We shall show that the requirements of the lemma are met.

We see from 2.3.6 that if $x \in H$ we have $\rho(x)W = W$. Conversely, if this is so, then $\rho(x)f_i \in I$ $(1 \leq i \leq r)$, whence $\rho(x)I \subset I$. It follows from 2.3.6 that $x \in H$. The proof of the Lie algebra property is quite similar, using 3.3.7 instead of 2.3.6.

Now let V be a finite dimensional vector space and W a subspace, of dimension d. The dth exterior power $\Lambda^d V$ contains the 1-dimensional subspace $L = \Lambda^d W$. If $x \in GL(V)$ let $\phi_d x$ be the induced linear map of $\Lambda^d V$. Then ϕ_d is a rational representation of $GL(V)$.

5.1.2. *Lemma*. (i) *Let* $x \in GL(V)$. *We have* $x.W = W$ *if and only if* $(\phi_d x)L = L$;

(ii) *Let* $X \in \mathfrak{gl}(V)$. *We have* $XW \subset W$ *if and only if* $d\phi_d(X)L \subset L$.

The "only if" parts are clear. Let $(v_1,\ldots,v_n)$ be a basis of V such that $(v_1,\ldots,v_d)$ is one of W. The elements $v_{i_1} \wedge \ldots \wedge v_{i_d}$ with $i_1 < \ldots < i_d$ form a basis of $\Lambda^d V$ and $v_1 \wedge \ldots \wedge v_d$ is a basis element of L.

Let $x \in$ ~~W~~ $GL(V)$. Choose the basis such that, moreover, $(v_{l+1},\ldots,v_{l+d})$ is a basis of xW, for some l. Put $e_1 = v_1 \wedge \ldots \wedge v_d$, $f = v_{l+1} \wedge \ldots \wedge v_{l+d}$. Then $(\phi_d x)e = f$. If $l > 0$ then e and f are linearly independent, and $\phi_d x$ cannot stabilize L. This implies (i).

If $X \in \mathfrak{gl}(V)$ we have, by 3.3.14(4)

$$d\phi_d(X)e = \sum_{i=1}^{d} v_1 \wedge \ldots \wedge v_{i-1} \wedge Xv_i \wedge v_{i+1} \wedge \ldots \wedge v_d.$$

Writing $Xv_i = \Sigma a_{ij} v_j$ it follows that

$$d\phi_d(X)e = \sum_{i=1}^{d} \sum_j a_{ij} v_1 \wedge \ldots \wedge v_{i-1} \wedge v_j \wedge v_{i+1} \wedge \ldots \wedge v_d,$$

from which we see that if $a_{ij} \neq 0$ for some $i \leq d$ and some $j > d$, the subspace L cannot be mapped into itself by $d\phi_d(X)$. This implies (ii).

5.1.3. *Theorem* (*Chevalley*). *Let* G *be a linear algebraic group and* H *a closed subgroup. There is a rational representation* $\phi: G \to GL(V)$ *and a* 1-*dimensional subspace* $L \subset V$ *such that*

$H = \{x \in G \mid (\phi x)L = L\}$,

$\mathfrak{h} = \{X \in \mathfrak{g} \mid d\phi(X)L \subset L\}$.

This is a consequence of the preceding lemmas.

We shall need below the following corollary of 5.1.3. Recall that a quasi-projective algebraic variety is an open subvariety of a projective variety (1.7.1).

5.1.4. *Corollary. Let G and H be as before. There is a quasi-projective homogeneous space X for G, together with a point x ∈ X such that*

(a) *the isotropy group of x in G is H;*

(b) *the morphism ψ: $g \mapsto g.x$ of G onto X defines a separable morphism $G^0 \to \psi G^0$;*

(c) *the fibers of ψ are the cosets gH (g ∈ G).*

Let V and L be as in 5.1.3. Let $\mathbb{P}(V)$ be the projective space defined by V. The points of $\mathbb{P}(V)$ are the 1-dimensional subspaces of V, let $x \in \mathbb{P}(V)$ be the point defined by L. We denote by π the canonical morphism $V-\{0\} \to \mathbb{P}(V)$. Now G acts on $\mathbb{P}(V)$ by $g.\pi(v) = \pi(\phi(g)v)$. Denote by X the G-orbit of x. By 4.3.1(i) (see also 4.3.5(1)) we have that X is a quasi-projective variety. (a) and (c) then follow from 5.1.3. That (b) holds is a consequence of the Lie algebra assertion of 5.1.3, using 3.2.21(ii) (check this).

5.2. Construction of quotients.

5.2.1. Let G and H be as before. A *quotient of G by H* is a

pair $(G/H,a)$ of a homogeneous space G/H for G, together with a point $a \in G/H$ whose isotropy group is H, such that the following universal property holds:
for any pair (Y,b) of a homogeneous space Y and a point $b \in Y$ whose isotropy group contains H, there is a unique G-morphism $\phi: G/H \to Y$ with $\phi a = b$.

The main result of this section is the following existence theorem.

5.2.2. Theorem. A quotient $(G/H,a)$ exists and is unique up to a G-isomorphism. In fact, if X and x are as in 5.1.4 then (X,x) is such a quotient.
The uniqueness part is trivial. To prove the existence statement we proceed as follows. We first define G/H as a ringed space. Using 5.1.4 it will be shown that this ringed space is an algebraic variety.
The set G/H is the set of cosets gH ($g \in G$), and a is the coset H. Let $\pi: G \to G/H$ be the canonical map.
We define $U \subset G/H$ to be open if and only if $\pi^{-1}U$ is open in G. This defines a topology on G/H, such that π is an open map (which is in accordance with 4.3.3(i)). We define a sheaf $\mathcal{O}$ of k-valued functions on G/H (see 1.4.2) as follows. Let $U \subset G/H$ be open. Then $\mathcal{O}(U)$ is the ring of functions f on U such that $f \circ \pi$ is regular on $\pi^{-1}U$ (check that this defines indeed a sheaf).
G acts transitively on G/H by left translations (set-theoretically), and these left translations are isomorphisms of the

ringed space $(G/H,\mathcal{O})$. More precisely, if $g \in G$, the map $(g,g'H) \mapsto gg'H$ is an isomorphism of $(G/H,\mathcal{O})$ (check this). It is then a straightforward matter to verify that if (Y,b) is as in 5.2.1, there is a unique G-morphism of ringed spaces $\phi: G/H \to Y$ which is G-equivariant and such that $\phi a = b$ (ϕ is given by $\phi(gH) = g.b$). To finish the proof of 5.2.2 it is sufficient to show that our ringed space G/H is isomorphic to a homogeneous space in the sense of 2.3.1. To do this we use 5.1.4. Let X,x and ψ be as in 5.1.4. By the properties of the ringed space G/H there is a G-equivariant morphism of ringed spaces $\phi: G/H \to X$ with $\phi a = x$. We shall show that ϕ is a isomorphism of ringed spaces, which will finish the proof.

First it follows from the definition of G/H and property (c) of 5.1.4 that ϕ is a continuous bijection. Let $U \subset G/H$ be open. Then $\phi U = \psi(\phi^{-1}U)$ is open because of 4.3.3(i). Consequently ϕ^{-1} is continuous, and ϕ is a homeomorphism of topological spaces.

To show that ϕ is an isomorphism of ringed spaces the following has to be established: if U is an open set in X, the homomorphism of k-algebras $\mathcal{O}_X(U) \to \mathcal{O}(\phi^{-1}U)$ defined by ϕ is an isomorphism. By the definition of $\mathcal{O}$ this means the following: for any regular function f on $V = \psi^{-1}U$ such that $f(gh) = f(g)$ $(g \in V, h \in H)$, there is a unique regular function F on U such that $F(\psi g) = f(g)$, if $g \in V$. To prove this we may assume, using 4.3.2, that G is connected. Let $\Gamma \subset V \times \mathbb{A}^1$ be the graph of f, i.e. $\Gamma = \{(g,f(g)) | g \in V\}$, let $\Gamma' = (\psi \times \mathrm{id})\Gamma$, so

$\Gamma' \subset U \times \mathbb{A}^1$. Then $U \times \mathbb{A}^1 - \Gamma' = (\psi \times id)(V \times \mathbb{A}^1 - \Gamma)$. Since Γ is closed in $V \times \mathbb{A}^1$ (1.6.10(i)) we see that $V \times \mathbb{A}^1 - \Gamma$ is open in $V \times \mathbb{A}^1$. It then follows from 4.3.3(i) that Γ' is closed in $U \times \mathbb{A}^1$. Let $\lambda: \Gamma' \to U$ be the morphism induced by projection on the first factor. It follows from the definitions that λ is bijective, and by property 5.1.4(b) (using 3.2.21) we have that λ is also separable. We conclude from 4.1.6 and 4.1.2 that there is an open set $U_1 \subset U$ such that λ induces an isomorphism of varieties $\lambda^{-1}U_1 \to U_1$. But if $u \in U$ there is $g \in G$ with $u \in g.U_1$ and the definitions show that the open set $g.U_1 \cap U \subset U$ has the same property as U_1. Hence λ is an isomorphism in a neighborhood of any point of Γ', consequently λ is an isomorphism $\Gamma' \to U$. This shows that there is a regular function F on U such that $\Gamma' = \{(u,F(u)) | u \in U\}$, and F is as required. This concludes the proof of 5.2.2.

5.2.3. *Corollary*. (i) *G/H is a quasi-projective variety, whose dimension is* dim G - dim H;
(ii) *If* G *is connected the morphism* $g \mapsto g.a$ *of* G *onto* G/H *is separable*.
These statements follow by 5.1.4 and 4.3.3.

5.2.4. *Exercises*. G is a linear algebraic group and H a closed subgroup.

(1) The following properties are equivalent: (a) G/H meets all components of G, (c) $G = G^0H$.

(2) An open subvariety of an affine algebraic variety is called a quasi-affine variety (an example of a quasi-affine

variety which is not affine is given in 1.6.12(4)). If all homomorphisms of algebraic groups $H \to \mathbb{G}_m$ are trivial then G/H is quasi-affine.

(3) Let $G = \mathbb{SL}_2$, $H \subset G$ the subgroup of upper triangular matrices. Then $G/H \simeq \mathbb{P}^1$, the projective line.

(4) Let H_i be a closed subgroup of the linear algebraic group G_i, $i = 1,2$. Show that there is an isomorphism of $G_1 \times G_2$-spaces $G_1 \times G_2/H_1 \times H_2 \simeq G_1/H_1 \times G_2/H_2$.

(5) Let H and K be connected closed subgroups of the linear algebraic group G.

(a) HK is an irreducible quasi-affine subvariety of G, whose dimension equals $\dim H + \dim K - \dim(H \cap K)$ (Hint: let $H \times K$ act on G by $(x,y)g = xgy^{-1}$ and consider the orbit of e).

(b) Let $\pi: G \to G/K$ be the morphism with $\pi g = gK$ (the points of G/K being identified with the cosets gK). By (a) we have $\dim \pi H = \dim H - \dim(H \cap K)$. Show that the restriction of π to H is a separable morphism onto its image if and only if $\dim(\mathrm{Lie}(H) \cap \mathrm{Lie}(K)) = \dim(H \cap K)$.

(c) Let $G = \mathbb{SL}_2$, H the subgroup of diagonal matrices and K a distinct conjugate of H. Show that, if char $k = 2$, we have $\dim(\mathrm{Lie}(H) \cap \mathrm{Lie}(K)) \neq \dim(H \cap K)$.

(6) Let char $k \neq 2$. Let $G = \mathbb{SO}_n$, $n \geq 3$ (see 2.1.3(4)), acting in $V = k^n$ in the standard manner. Put $X = \{(x_1,\dots,x_n) \in V \mid x_1^2 + \dots + x_n^2 = 1\}$, $e = (0,\dots,0,1)$. Then G acts on X. The isotropy group of $e \in X$ is $\mathbb{SO}_{n-1}$, viewed as a subgroup of $\mathbb{SO}_n$. Show that $\mathbb{SO}_n/\mathbb{SO}_{n-1}$ is isomorphic to the affine G-variety X. (Hint: use Witt's theorem, see [26, Ch. XIV, §5]).

We next discuss the case of a normal subgroup.

5.2.5. <u>Proposition</u>. <u>Let</u> H <u>be a closed normal subgroup of the linear algebraic group</u>.

(i) G/H <u>is an affine variety</u>;

(ii) <u>Provided with the usual group structure</u>, G/H <u>becomes a linear algebraic group</u>.

Make G/H into a homogeneous space for $G \times G$ by defining $(g_1,g_2).xH = g_1xg_2^{-1}H$. The isotropy group of H contains $H \times H$. Using the universal property of 5.2.1 and 5.2.4(4) we see that the map $(gH,g_1H) \mapsto gg_1^{-1}H$ defines a morphism of varieties $G/H \times G/H \to G/H$. It readily follows that G/H is an algebraic group. To finish the proof of 5.2.5 it remains to prove (i). For this we use 5.1.3. Let V,ϕ and L be as in 5.1.3. For any rational character χ of H put

$$V_\chi = \{v \in V | \phi(h)v = \chi(h)v, h \in H\}.$$

From the linear independence of characters it follows that the subspace V' of V spanned by the V_χ is the direct sum of the nonzero V_χ. Since H is normal, G permutes the V_χ. It follows that the subspace V' spanned by the V_χ is stable under ϕG. We may suppose that $V' = V$.

Let $W \subset \mathrm{End}(V)$ be the space of linear transformations stabilizing each V_χ. Define a rational map $\psi: G \to GL(W)$ by

$$\psi(x)f = \phi(x)f\phi(x)^{-1} \quad (x \in G, f \in W).$$

(Check that $\psi(x)f \in W$ if $f \in W$).

If $\psi(x) = \mathrm{id}$, we have that $\phi(x)$ commutes with all $f \in W$, and

$\phi(x)$ must act as a scalar multiplication in each V_χ. In particular, $\phi(x)$ stabilizes L and $x \in H$. Hence ψ induces an injective map $\lambda: G/H \to GL(W)$, which is, in fact, a homomorphism of algebraic groups (check this). The image of λ is a closed subgroup of GL(W) (2.2.5(ii)). We shall prove that λ is an isomorphism of G/H into that image, which will establish (i). By 4.3.3(iii) it suffices to show that the tangent map $(d\lambda)_e$ is injective. This amounts to proving the following: if $X \in$ Lie G and $(d\psi)_e X = 0$, then $X \in$ Lie H. Now it follows from 3.3.14(1) that $(d\psi)_e X(f) = (d\phi)_e(X)f - f(d\phi)_e(X)$. Hence $(d\psi)_e X = 0$ implies that $(d\phi)_e(X)$ commutes with all $f \in W$, and it again follows that $(d\phi)_e(X)L \subset L$, whence $X \in$ Lie H by 5.1.3.

5.2.6. Exercises. H is a closed subgroup of the linear algebraic group G.

(1) Let N be a closed normal subgroup contained in H. Then $G/H \simeq (G/N)/(H/N)$.

(2) Let $A = \{f \in k[G] \mid f(gh) = f(g),\ g \in G, h \in H\}$.

(a) If G/H is affine the $k[G/H] = A$.

(b) For G/H to be affine it is necessary and sufficient that A is a k-algebra of finite type which separates the cosets of H. The latter requirement means that if xH and yH are distinct cosets there is $f \in A$ such that $f(xH) = 1$, $f(yH) = 0$. (Hint: If A is a k-algebra of finite tvpe let X be an affine algebraic variety with $k[X] = A$. Show that left translations induce a structure of homogeneous space on X, which satisfies the requirements of 5.2.1).

(c) Let H be a torus. The G/H is affine (Hint: If xH and yH are distinct cosets, let I and J be the ideals in k[G] which they define. Then I+J = k[G]. Deduce from 2.5.2(c) that $I \cap A + J \cap A = A$ and verify the condition of (b)).

Notes.

The discussion of quotients in this chapter is similar to the one in [3 ,§6] or [23 , no.12].

By exploiting homogeneity, we could avoid the use of a result of Chevalley [23 , p.41,Th.B], which depends on normality.

6. Solvable groups.

This chapter deals with the structure theory of connected solvable linear algebraic groups, which is basic for the general theory. We start with some general facts.

6.1. Proposition. *Let* G *be an algebraic group. Let* $(X_i, f_i)_{i \in I}$ *be a family of irreducible varieties, together with morphisms* $f_i : X_i \to G$. *Assume that* $e \in Y_i = f_i X_i$ $(i \in I)$. *Let* H *be the smallest closed subgroup of* G *containing all* Y_i.

(i) H *is connected*;

(ii) *There is an integer* $n \geq 0$, *a set* $a = (a(1), \ldots, a(n)) \in I^n$ *and signs* $\varepsilon(i) = \pm 1$ $(i \in I)$ *such that* $H = Y_{a(1)}^{\varepsilon(1)} \ldots Y_{a(n)}^{\varepsilon(n)}$.

We may assume that the sets Y_i^{-1} occur among the Y_j. For each $a = (a(1), \ldots, a(n)) \in I^n$ (n some integer ≥ 0) we write $Y_a = Y_{a(1)} \ldots Y_{a(n)}$. Since Y_a is the image under a morphism of the irreducible variety $X_{a(1)} \times \ldots \times X_{a(n)}$, it is an irreducible subset of G, and the same is true for its closure $\overline{Y}_a$ (1.2.3). We have $Y_b . Y_c = Y_{(b,c)}$, where (b,c) denotes the concatenation of b and c. This implies that $\overline{Y_b . Y_c} \subset \overline{Y}_{(b,c)}$ (use an argument as in the proof of 2.2.4). Now choose a such that $\dim \overline{Y}_a$ is maximal. For any b we have $\overline{Y}_a \subset \overline{Y}_a . \overline{Y}_b \subset \overline{Y}_{(a,b)}$, whence $\overline{Y}_{(a,b)} = \overline{Y}_a$ for all b, and $\overline{Y}_b \subset \overline{Y}_a$. It also follows that $\overline{Y}_a . \overline{Y}_a = \overline{Y}_a$, from which one concludes that $\overline{Y}_a$ is a group. Since Y_a contains an open subset of $\overline{Y}_a$ (1.9.5) it follows that $\overline{Y}_a = Y_a . Y_a$ (2.2.3). This implies 6.1.

6.2. Corollary. *Let* G *be an algebraic group. Assume that there is a family of closed connected subgroups* $(H_i)_{i \in I}$ *such that*

G is generated by them, as an abstract group. Then G is connected and there is an integer $n \geq 0$ *and a set* $(a(1),\dots,a(n)) \in I^n$ *such that* $G = H_{a(1)} \dots H_{a(n)}$.

If H and K are subgroups of the group G we denote by (H,K) the subgroup generated by the commutators $(x,y) = xyx^{-1}y^{-1}$ $(x \in H,\ y \in K)$.

6.3. *Corollary. Let G be an algebraic group, and H,K closed subgroups, one of which is connected. Then (H,K) is closed and connected.*

Assume, for example, that H is connected. Then 6.3 follows from 6.1, applied in the situation that I = K, all X_i are H and $f_y x = xyx^{-1}y^{-1}$ $(y \in K)$.

6.4. Recall that if G is an abstract group, the *higher commutator groups* $\mathcal{D}^n G$ are defined by $\mathcal{D}^0 G = G$, $\mathcal{D}^{n+1} G = (\mathcal{D}^n G, \mathcal{D}^n G)$ and that G is solvable if $\mathcal{D}^n G = \{e\}$ for some n. In that case, $\mathcal{D}^1 G \neq G$ and $G/\mathcal{D}^1 G$ is abelian.

Define the normal subgroups $\mathcal{C}^n G$ by $\mathcal{C}^0 G = G$, $\mathcal{C}^{n+1} G = (G, \mathcal{C}^n G)$. Then G is *nilpotent* if $\mathcal{C}^n G = \{e\}$ for some n, which is tantamount to saying that all repeated commutators $(x_1(\dots(x_n, x_{n+1})\dots)$ equal e. The *center* Z(G) is the subgroup $\{x \mid (x,G) = e\}$.

6.5. *Corollary. Let G be a connected algebraic group.*

(i) The groups $\mathcal{D}^n G$ *and* $\mathcal{C}^n G$ *are closed connected subgroups;*

(ii) Z(G) is a closed subgroup. If G is nilpotent and $\neq \{e\}$, *the identity component of Z(G) is non-trivial.*

(i) follows from 6.3. It is obvious that $Z(G)$ is closed. If G is nilpotent and $\neq \{e\}$, let n be such that $C^nG \neq \{e\}$, $C^{n+1}G = \{e\}$. Then C^nG is connected, non-trivial, and lies in $Z(G)$.

6.6. Exercises. (1) Assume char $k \neq 2$. Let $V = k^n$ and define a symmetric bilinear form $\langle\ ,\ \rangle$ on V by

$$\langle (x_1,\ldots,x_n),(y_1,\ldots,y_n)\rangle = \sum_{i=1}^{n} x_i y_i .$$

Then $\mathbb{O}_n$ (see 2.1.3(4)) is the group of linear transformations $g \in \mathbb{GL}_n$ such that $\langle gx,gy\rangle = \langle x,y\rangle$ for all $x,y \in V$.
If $x \in V$, $\langle x,x\rangle = 1$ put $s_x y = y - \langle x,y\rangle x$. This is a reflection, contained in $\mathbb{O}_n$ and one knows that the s_x generate $\mathbb{O}_n$ (see [25 , p.353]). Put $X = \{(x,y) \in V \times V \mid \langle x,x\rangle = \langle y,y\rangle = 1\}$, this is an irreducible affine variety. Then $f: (x,y) \mapsto s_x s_y$ defines a morphism $X \to \mathbb{SO}_n$.
(a) Deduce from 6.1 that $\mathbb{SO}_n$ is connected.
(b) Show that the condition $e \in Y_i$ of 6.1 cannot be omitted.
(2) Let $V = k^{2n}$ and define an alternating bilinear form $(\ ,\ )$ on V by $((x_1,\ldots,x_{2n}),(y_1,\ldots,y_{2n})) = \sum_{i=1}^{n} (x_i y_{n+i} - x_{n+i} y_i)$. The group of linear transformations $g \in \mathbb{GL}_n$ such that $(gx,gy) = (x,y)$, for all $x,y \in G$, is the symplectic group $\mathbb{Sp}_{2n}$ (see 2.1.3(4)). One knows that $\mathbb{Sp}_{2n}$ is generated by the symplectic transvections $t_{x,a}$ $(x \in V, a \in k)$ with $t_{x,a}y = y - a(x,y)x$ (see [25 , p. 373]). Deduce from 6.1 that $\mathbb{Sp}_{2n}$ is connected.
(3) Let G be a connected nilpotent linear algebraic group and H a proper connected closed subgroup. Show that $\dim H <$ $< \dim N_G H$, where $N_G H$ is the normalizer. In particular, if $\dim H = \dim G - 1$ then H is a normal subgroup (Hint: use

6.5(ii)).

(4) The group $\mathbb{U}_n$ of unipotent upper triangular matrices (2.1.3(4)) is connected and nilpotent.

Until 6.14 we assume that G is <u>a connected solvable linear algebraic group</u>.

6.7. <u>Theorem</u> (<u>Lie-Kolchin</u>). <u>Let</u> G <u>be a connected solvable closed subgroup of</u> $\mathbb{GL}_n$. <u>There is</u> $x \in \mathbb{GL}_n$ <u>such that</u> xGx^{-1} <u>lies in the subgroup</u> $\mathbb{T}_n$ <u>of upper triangular matrices</u>.

Let $V = k^n$, then G operates in V. It suffices to prove (check this) that there is a nonzero $v \in V$ which is a common eigenvector of all $g \in G$. We prove this by induction on dim G, starting with $G = \{e\}$. By 6.5(i), the commutator subgroup $(G,G) = \mathcal{D}^1 G$ is a connected solvable closed subgroup, which has smaller dimension. By induction we may assume there is a common eigenvector for the elements of (G,G). If χ is a rational character of (G,G), let $V_\chi = \{v \in V | g.v = \chi(g)v, g \in (G,G)\}$. Then G permutes the distinct nonzero subspaces V_χ. Since G is connected, it must stabilize all these V_χ. We may then, to prove the existence of a common eigenvector for the elements of G, assume that $V = V_\chi$ for some χ. The elements of (G,G) are thus scalar multiplications. But since a commutator $xyx^{-1}y^{-1}$ $(x,y \in \mathbb{GL}_n)$ has determinant 1, these scalar multiplications also have determinant 1, which implies that (G,G) is finite. Being connected, it must be trivial, and we are in the case that G is abelian. In that case the result follows from 2.4.2(i).

6.8. Corollary. *Let G be a connected nilpotent linear algebraic group.*

(i) *The sets* G_s, G_u *of semi-simple and unipotent elements are closed connected subgroups, and* G_s *is a central torus of* G;

(ii) *The product map* $\pi: G_s \times G_u \to G$ *is an isomorphism of algebraic groups.*

We know by 4.4.8 that G_s is a central subgroup of G. Assume that G is a closed subgroup of $\mathbb{GL}_n$. From 2.4.2(ii) we obtain a G-invariant direct sum decomposition of k^n, such that the elements of G_s act as scalar multiplications in all summands. By 6.7 we can trigonalize the restriction of G to any summand. The proof of 6.8 then proceeds as that of 2.4.12.

6.9. Corollary. *Let G be a connected solvable linear algebraic group.*

(i) (G,G) *is a closed unipotent, connected subgroup.*

(ii) *The set* G_u *of unipotent elements is a closed nilpotent, connected, normal subgroup of* G. *The quotient group* G/G_u *is a torus.*

Using 6.7 we may assume that G is a closed subgroup of $\mathbb{T}_n$. It is then clear that (G,G) is unipotent. We know already that (G,G) is connected and closed. This establishes (i).

Since $G_u = G \cap \mathbb{U}_n$ (where $\mathbb{U}_n$ is the group of unipotent upper triangular matrices) it is clear that G_u is a closed normal subgroup, which is nilpotent because $\mathbb{U}_n$ is nilpotent (6.6(4)). Also there is an injective homomorphism of algebraic groups $G/G_u \to \mathbb{T}_n/\mathbb{U}_n$. Since the last group is a torus, all elements of G/G_u must be semi-simple. Being connected, this group is a torus.

It remains to be seen that G_u is connected. Its identity component G_u^0 is a closed normal subgroup of G. Replacing G by the quotient group G/G_u^0 we are reduced to showing: if G_u is finite, it is trivial. Now if G_u is finite, it must lie in the center of G (because of the connectedness of G), and it follows that G is nilpotent. But then we see from 6.8(ii) that G_u is connected, hence must be trivial.

The next lemma is an essential tool in the proof of 6.11.

6.10. Lemma. Assume G is connected solvable but not a torus. There is a closed normal subgroup N of G, which is contained in the center of G_u, and is isomorphic to the additive group $\mathbb{G}_a$.

Let H be a non-trivial connected closed normal subgroup of G which is contained in the center of G_u. Such a group exists, as follows from the proof of 6.5(ii). If char k = p > 0, we may, moreover, assume that $H^p = \{e\}$ (replace H by its image under a suitable map $x \mapsto x^{p^e}$). Then, by 2.6.7(1), H is isomorphic to a vector group $(\mathbb{G}_a)^n$. If n = 1 we are done. Otherwise, let $V \subset k[H]$ be the subspace of additive functions (see 2.6). The torus $T = G/G_u$ operates on H via conjugation, hence it operates on k[H], stabilizing V. By 2.3.4(i) and 2.5.2 we can find a nonzero $\alpha \in V$ such that the line $k\alpha \subset V$ is stabilized by T. Now $(\ker \alpha)^0$ is an (n-1)-dimensional subgroup of H which has the same properties as H. The result follows by recurrence.

The next theorem gives the most important facts about the internal structure of a connected solvable linear algebraic group G. A _maximal torus_ of G is a subtorus which has the same dimension as the torus $S = G/G_u$. It will follow (see 6.12(i)) that a maximal torus is also a maximal torus in the set-theoretical sense.

6.11. _Theorem_. _Let_ G _be a connected solvable linear algebraic group_.

(i) _Let_ $s \in G$ _be semi-simple. Then_ s _lies in a maximal torus. In particular: maximal tori exist_;

(ii) _The centralizer_ $Z_G(s)$ _of_ s _is connected_;

(iii) _Two maximal tori of_ G _are conjugate_;

(iv) _If_ T _is any maximal torus, the product map_ $\psi: T \times G_u \to G$ _is an isomorphism of varieties_.

We first prove (iv). Let the direct product $T \times G_u$ operate on G by $(t,g)x = txg^{-1}$. The orbit of e is the subgroup $T.G_u$. Since the isotropy group of e in $T \times G_u$ is trivial (because $T \cap G_u = \{e\}$) we have $\dim \overline{T.G_u} = \dim T + \dim G_u = \dim G$. It follows from 4.3.1(i) and 2.2.4(ii) that $G = T.G_u$. Hence G is a homogeneous space for $T \times G_u$. From 3.3.11 one sees that the tangent map $(d\psi)_{(e,e)}$ is the map $(X,Y) \mapsto X-Y$ of $L(T) \times L(G_u)$ to $L(G)$. Since $L(T) \cap L(G_u) = \{0\}$ (by 3.4.3(2),(3)) this map is injective, hence bijective, and 4.3.3(iii) implies that ψ is an isomorphism of varieties, proving (iv).

We first prove the other assertions in the case that $\dim G_u = 1$. Since G_u is connected (6.9(ii)) we know from 2.6.6 that $G_u \cong \mathbb{G}_a$. Fix an isomorphism $\phi: \mathbb{G}_a \cong G_u$. Let $\pi: G \to S = G/G_u$ be the canonical homomorphism. There is a rational character α of S such that

$$(*) \quad g\phi(\xi)g^{-1} = \phi(\alpha(\pi g)\xi), \ g \in G, \ \xi \in k.$$

(this follows, for example, from 4.1.8(1)).
If $\alpha = 1$, then G is commutative, and all assertions follow from 2.4.12. So assume $\alpha \neq 1$.
Let $s \in G$ be semi-simple and put $Z = Z_G(s)$. By 4.4.5(ii) we have a direct sum decomposition $\mathfrak{g} = (\mathrm{Ad}(s)-1)\mathfrak{g} \oplus \mathfrak{z}$. Since $\pi(sxs^{-1}) = \pi x$ we have $d\pi \circ (\mathrm{Ad}(s)-1)\mathfrak{g} = 0$, so $(\mathrm{Ad}(s)-1)\mathfrak{g} \subset \mathrm{Ker}\, d\pi = \mathrm{Lie}\, G_u$ (because, by 5.2.3(ii), π is a separable homomorphism, with kernel G_u). It follows that $\dim(\mathrm{Ad}(s)-1)\mathfrak{g} \leqslant 1$, $\dim Z^0 = \dim \mathfrak{z} \geqslant \dim G - 1$. Let s be such that $\alpha(\pi s) \neq 1$. Such s exist: take the semi-simple part of some g with $\alpha(\pi g) \neq 1$. Then (*) shows that $Z \cap G_u = \{e\}$. Hence $Z^0 \neq G$, and $\mathfrak{z} \neq \mathfrak{g}$. It follows that Z^0 is a closed connected, solvable subgroup of G, with $(Z^0)_u = \{e\}$, of dimension n-1. By 6.9(ii), Z^0 is a maximal torus. By (iv) we have $G = Z^0G_u$. If $g = xy$ ($x \in Z^0$, $y \in G_u$) commutes with s, then y commutes with s, hence $y = e$, whence $Z = Z^0$. We have now shown that the centralizer of a semi-simple element $s \in G$ with $\alpha(\pi s) \neq 1$ is a maximal torus. If $\alpha(\pi s) = 1$, then a similar argument shows that s lies in the center of G. Hence s commutes with all elements of G, which implies that it lies in a maximal torus (viz. the centralizer of some s' with $\alpha(\pi s') \neq 1$).
It remains to prove (iii) (still in the case $\dim G_u = 1$). Now

if T is some maximal torus, π induces a bijection $T \to S$. It follows that, if $t \in T$ is such that $\alpha(\pi t) = 1$, we have $T = Z_G(t)$. Let T' be another maximal torus, and choose $t' \in T'$ such that $T' = Z_G(t')$. We write $t' = t\phi(\xi)$ $(t \in T, \xi \in k)$. From (*) we deduce that

$$\phi(\eta)t'\phi(\eta)^{-1} = t\phi(\xi+(\alpha(\pi t)^{-1}-1)\eta).$$

Since $\alpha(\pi t) = \alpha(\pi t') \neq 1$ we can find $\eta \in k$ such that $\phi(\eta)t'\phi(\eta)^{-1} = t$. It follows that $\phi(\eta)T'\phi(\eta)^{-1} = T$, whence (iii).

Next consider the general case. Assume $\dim G_u > 1$. Let N be as in 6.10. We put $\overline{G} = G/N$. Then $\dim G/G_u = \dim \overline{G}/\overline{G}_u$. Let $s \in G$ be semi-simple, and let $\overline{s}$ be its image in $\overline{G}$. By an induction on dim G we may assume that $\overline{s}$ lies in a maximal torus $\overline{T}$ of $\overline{G}$. The inverse image of $\overline{T}$ in G is a connected closed subgroup H of G (see 5.2.4(1)), and $\dim H_u \leq 1$. We know already that s lies in a maximal torus of H, which is also one of G. This proves (i). The proof of (iii) proceeds similarly, and can be left to the reader.
Let $G_1 = \{g \in G \mid sgs^{-1}g^{-1} \in N\}$. The connectedness of $Z_{\overline{G}}(\overline{s})$, which we may assume, implies that G_1 is a connected closed subgroup of G, containing $Z_G(s)$ and N, and such that $G_1/N \simeq Z_{\overline{G}}(\overline{s})$. If $G_1 \neq G$ we have, by induction on dim G, that $Z_G(s)$ is connected. So assume $G_1 = G$. Put $Z = Z_G(s)$. We have again a direct sum decomposition $\mathfrak{g} = \mathfrak{z} \otimes (\mathrm{Ad}(s)-1)\mathfrak{g}$. Proceeding as in the first part of this proof, one shows that $(\mathrm{Ad}(s)-1)\mathfrak{g} \subset \mathrm{Lie}\, N$. If $(\mathrm{Ad}(s)-1)\mathfrak{g} = \{0\}$, then s is central in

G and $Z_G(s) = G$. If $(\mathrm{Ad}(s)-1)\mathfrak{g} = N$, then $\dim Z^0 = \dim G - 1$ and we have $Z \cap N = \{e\}$ (use a relation like (*)). Then the product map $Z^0 \times N \to G$ is bijective, from which it follows again that $Z = Z^0$. This finishes the proof of the theorem.

6.12. Corollary. *Let $H \subset G$ be a subgroup of G all of whose elements are semi-simple.*
(i) *H lies in a maximal torus of G. In particular, any subtorus of G lies in a maximal torus;*
(ii) *The centralizer Z_GH is connected and coincides with the normalizer N_GH.*
If $s \in G$ is semi-simple and contained in the center Z(G) of G, it lies in all maximal tori of G (by 6.11(i) and (iii)). This implies (i) in the case that $H \subset Z(G)$, and then (ii) is obvious.
If $H \not\subset Z(G)$, choose $s \in H - Z(G)$. Since the restriction of the quotient map $G \to G/G_u$ to H is bijective, it follows that H is commutative. So $H \subset Z_G(s)$. Then 6.11(ii) implies (i), by induction on dim G.
The same argument shows that Z_GH is connected. Finally, if $x \in N_GH$ then for all $h \in H$ we have $xhx^{-1}h^{-1} \in H \cap (G,G) \subset$ $\subset H \cap G_u = \{e\}$, whence $N_GH = Z_GH$.

6.13. Corollary. *Let H be a proper closed subgroup of G. There exists a non-constant regular function on the variety G/H.*

We use induction on dim G + dim G/H. This integer is at least 2, and if it equals 2 then dim G = dim G/H = 1. Then

G ($\simeq \mathbb{G}_a, \mathbb{G}_m$) is abelian, H is a normal subgroup and G/H is affine (5.2.5(i)). This implies the assertion in that case. If $G_u = \{e\}$ than G is again abelian, and G/H is affine by the same argument. If $G_u \neq \{e\}$, let N be as in 6.10. Then HN is a closed subgroup. If it is a proper one, we may assume by induction that there is a non-constant regular function on G/HN, and its pull-back to G/H (via the morphism $G/H \to G/HN$) is one on G/H. If HN = G, the group $(H^0)_u$ is a normal subgroup of G, and $G/H \simeq G/(H^0)_u/H/(H^0)_u$. If $(H^0)_u \neq \{e\}$ we can again use induction. If $(H^0)_u = \{e\}$ then H^0 is a torus T, which must be a maximal torus of G. Moreover, now dim $G_u = 1$. It then follows from 6.11(iv) that G/H and G_u/H_u are isomorphic, whence the assertion (since dim $G_u = 1$).

6.14. Let G be an arbitrary linear algebraic group. Using 6.2 one sees that the family of all closed, connected, solvable, normal subgroups generates a subgroup of the same kind. Hence there is a maximal closed, connected, solvable normal subgroup, the _radical_ RG of G. Similarly, there is a maximal closed, connected, unipotent, normal subgroup of G, the _unipotent radical_ R_uG. We have $R_uG = (RG)_u$ (check this).
G is called _semi-simple_ if RG = {e} and _reductive_ if $R_uG = \{e\}$.

6.15. _Propostion. Let_ G _be a connected, reductive linear algebraic group. Then_ RG _is a torus. It coincides with the identity component of the center_ ZG. _Moreover_, RG ∩ (G,G) _is finite_.

Since $R_uG = \{e\}$, we have that RG is a torus T. Now G = $(N_GT)^0 = (Z_GT)^0$ (see 2.5.11), and T lies in the center of G. Since $R_uG = \{e\}$, we have $(ZG)^0_u = \{e\}$, hence $T = (ZG)^0$ by 2.4.12.

To prove the last statement, assume G to be a closed subgroup of some GL(V). Decompose V into a direct sum $\oplus V_\chi$, such that $t \in T$ acts as scalar multiplication by $\chi(t)$ in V_χ, the χ being distinct rational characters of T. Then $G \subset Z_{GL(V)}T$ stabilizes all V_χ, and $(G,G) \subset \prod_\chi SL(V_\chi)$, from which the last statement follows.

6.16. Exercises. (1) Give an example of a finite solvable subgroup of $\mathbb{GL}_2(\mathbb{C})$ which is not conjugate to a group of triangular matrices (this shows that the connectedness assumption in 6.6 is essential).

In the next exercises G is a connected solvable linear algebraic group.

(2) Let H be a closed, connected, nilpotent subgroup of G which is equal to its normalizer. Show that H is the centralizer of a maximal torus of G (Hint: By 6.8 we have $H = H_s \times H_u$. Show that $H = Z(H_s)$ and use 6.12(i) to prove that H_s is a maximal torus).

(3) There is a sequence $G = G_n \supset G_{n-1} \supset \ldots \supset G_1 \supset G_0 = \{e\}$ of closed, connected, normal subgroups of G such that all quotient groups G_i/G_{i-1} are either isomorphic to $\mathbb{G}_a$ or to $\mathbb{G}_m$.

(4) Let H be a closed subgroup of G such that dim G/H = 1. Then G/H is either isomorphic to $\mathbb{G}_a$ or to $\mathbb{G}_m$ (as a variety). (Hint: use 6.10).

Notes.

The first results of this chapter give standard elementary material. 6.3 is also true if H and K are closed subgroups (not necessarily connected), such that H normalizes K (see [3 , p.108].

The proof of the main result 6.11 is somewhat different from the usual one, which is due to Borel ([2 ,§12], see also [3 , p.244]). We have exploited here the auxiliary result 6.10.

6.13 is rather a weak result. As a matter of fact, in that case the quotient G/H is an affine variety (see [5, 9.13]).

7. <u>Complete varieties. Borel subgroups, parabolic subgroups and their properties</u>.

7.1. <u>Complete varieties</u>.

An algebraic variety X is said to be *complete* if the following holds: for any variety Y, the projection morphism $X \times Y \to Y$ is closed, i.e. maps closed sets onto closed sets. The example in 1.9.1 shows that the affine line $\mathbb{A}^1$ is not complete. One should view the notion of completeness as an analogue, in the category of algebraic varieties, of the notion of compactness in the category of locally compact topological spaces (see exercise 7.1.5(1)).

In the next proposition X and Y are varieties.

7.1.1. *Proposition*. *Let X be complete*.

(i) *A closed subvariety of X is complete*;

(ii) *If Y is complete then so is* $X \times Y$;

(iii) *If* $\phi: X \to Y$ *is a morphism, then* ϕX *is closed and complete*;

(iv) *If X is a subvariety of Y then X is closed in Y*;

(v) *If X is irreducible then any regular function on X is constant*;

(vi) *If X is affine then X is finite*.

(i) and (ii) follow from the definition of completeness in a straightforward manner. To prove (iii), let $\Gamma = \{(x,\phi(x)) \in X \times Y\}$ be the graph of ϕ. It is a closed subset of $X \times Y$ (1.6.10(i)), and ϕX is the projection of Γ onto Y. Since X is complete, this projection is closed. To prove that ϕX is also complete one uses that Γ is complete, since it is isomorphic to X. Then (iv) is a consequence of (iii), applied to the injection $X \to Y$. A regular function f on X can be viewed as a morphism $f: X \to \mathbb{A}^1$, which is also a morphism of X into the projective line $\mathbb{P}^1$. If X is irreducible and f non-constant, fX is a non-empty dense subset of $\mathbb{P}^1$ and then (iv) shows that $fX = \mathbb{P}^1$, which is impossible. This establishes (v). Finally, (vi) is a direct consequence of (v).

The following theorem is the main result of this section. Projective varieties were defined in 1.7.1.

7.1.2. Theorem. A projective variety is complete.

It suffices to prove that projective n-space $\mathbb{P}^n$ is complete (7.1.1(i)). So we have to prove that the projection morphism $\pi: \mathbb{P}^n \times Y \to Y$ is closed for all varieties Y. It suffices to assume Y affine and irreducible. Let $A = k[Y]$, $S = A[T_0, \ldots, T_n]$. Define the notion of a homogeneous ideal of S as in 1.7.3. We can view S as an algebra of functions on $k^{n+1} \times Y$. If I is a proper homogeneous ideal in S, define the subset $V^*(I)$ of $\mathbb{P}^n \times Y$ by

$$V^*(I) = \{(x^*, y) \in \mathbb{P}^n \times Y \mid f(x,y) = 0 \text{ for all } f \in I\}.$$

Here x^* denotes, as in 1.7.1, the point of $\mathbb{P}^n$ defined by $x \in k^{n+1} - \{0\}$. Results like those of 1.7.4 and 1.7.5(2) also hold in the present situation: the closed subsets of $\mathbb{P}^n \times X$ coincide with the $V^*(I)$; $V^*(I) = \emptyset$ if and only if there is $h > 0$ such that $T_i^h \subset I$ $(0 \leqslant i \leqslant n)$; $V^*(I)$ is irreducible if and only if I is a prime ideal.

We have to show that all sets $\pi V^*(I)$ are closed. We may assume I to be a prime ideal. We may also, replacing Y by $\overline{\pi V^*(I)}$, assume that $A \cap I = 0$.

So we have to prove: if $y \in Y$ there is $x^* \in \mathbb{P}^n$ such that $(x^*,y) \in V^*(I)$. Let $M \subset A$ be the maximal ideal in A defined by y. Then MS + I is also a homogeneous ideal, and what we must prove is equivalent to the statement that $V^*(MS+I) \neq \phi$. Assume $V^*(MS+1) = \phi$. Then $T_i^h \subset MS+I$ for some $h > 0$ and for all i. This implies that there is $\ell > 0$ such that the set $S_\ell \subset S$ of homogeneous polynomials of degree ℓ is contained in MS+I. Writing $N = S_\ell/S_\ell \cap I$, we have $N = MN$. We claim that this implies $N = 0$ (This is a version of Nakayama's lemma, see [26 , Ch.IX,§1]). Now N is a finitely generated A-module, let $(n_1,\ldots,n_r)$ be a set of generators. There are $m_{ij} \in M$ such that

$$n_i = \sum_{j=1}^{r} m_{ij}n_j.$$

If $a = \det(\delta_{ij}-m_{ij})$, it follows that $an_i = 0$. Also $a-1 \in M$, so $a \neq 0$. Let $f_i \in S_1$ represent n_i, then $af_i \in I$. Since S/I is an integral domain and $a \notin I$ (because $a \neq 0$ and $A \cap I = 0$), it follows that $f_i \in I$ and $n_i = 0$. Hence $N = 0$, establishing the claim. So $S_\ell \subset I$. But then $T_i^\ell \in I$ for all i and $\pi V^*(I)=\phi$, a contradiction. This proves the theorem.

The next result will only be used in the proof of 7.1.4, which does not concern the theory of linear algebraic groups.

7.1.3. Lemma. *Let X, Y, Z be irreducible varieties, with X complete. Assume that ϕ: X × Y → Z is a morphism and put $\phi_y(x) = \phi(x,y)$ ($x \in X$, $y \in Y$). If there is $a \in Y$ such that ϕ_a is constant, then ϕ_y is constant for all $y \in Y$.*

Let $\Gamma = \{(x,y, \phi(x,y)) \in X \times Y \times Z\}$ be the graph of ϕ, it is closed in X × Y × Z. Because of the completeness of X, the subset $A = \{(y,\phi(x,y)) \mid x \in X, y \in Y\}$ if Y × Z is closed, and also irreducible. Let π: A → Y be the projection. Then $\pi^{-1}a$ consists of only one point, and by 4.2.5 we have dim A=dim Y. If $x \in X$, the subset $A_x = \{(y,\phi(x,y)) | y \in Y\}$ of A is closed, being the graph of a morphism Y → Z, and is isomorphic to Y. Hence it is irreducible, and its dimension equals dim A. But then $A_x = A$, which means that ϕ_y is constant for all y.

7.1.4. Proposition. *Let G be a connected algebraic group, which is complete. Then G is commutative.*

This follows by applying 7.1.3 with X = Y = Z = G, $\phi(x,y) = xyx^{-1}$ and a = e.

The groups with the properties of 7.1.4 are called *abelian varieties*. An example are the elliptic curves, briefly mentioned in 2.1.3(5). For the theory of abelian varieties we refer to [29].

7.1.5. Exercises. (1) Let X be a locally compact Hausdorff space. Show that X is compact if and only if for any locally compact space Y the projection morphism X × Y → Y is closed (Hint: in the proof of the if-part use a one point compactification of a non-compact X, i.e. a compact space $\hat{X}$ containing X as a subspace and such that $\hat{X}$-X consists of one point).

(2) (a) Let G be a connected algebraic group and H a closed, connected, complete subgroup. Then H lies in the center of G.
(b) Let G be a connected algebraic group. There exists a maximal closed, connected, complete subgroup of G. It lies in the center of G.

7.2. Parabolic subgroups and Borel subgroups of a linear algebraic group.

Let G be a linear algebraic group.

7.2.1. Lemma. *If X and Y are homogeneous spaces of G and ϕ: X → Y a bijective G-morphism, then X is complete if and only if Y is so.*
It follows from 4.3.3(i) that for any variety Z the map $\phi \times id$: $X \times Z \to Y \times Z$ is a homeomorphism of topological spaces.
7.2.1 is a consequence of this fact.

A closed subgroup P of G is said to be a *parabolic subgroup* if the quotient variety G/P is complete. It follows readily from the definition that P is parabolic in G if and only if $P \cap G^0$ is parabolic in G^0.

7.2.2. Proposition. *Assume G to be connected. There exist proper parabolic subgroups if and only if G is non-solvable.*
If G is solvable it follows from 6.13 and 7.1.1(v) that proper parabolic subgroups do not exist.
Assume that G is a closed subgroup of some $\mathbb{GL}(V)$. Then G operates on the projective space $\mathbb{P}(V)$ and by 4.3.1(ii) there is a closed orbit X for this action. By 7.1.2, X is a complete

homogeneous space for G. Let $x \in X$, and let P be the isotropy group of x in G. Then $gP \mapsto g.x$ defines a bijective G-morphism of homogeneous spaces $G/P \to X$ and 7.2.1 shows that P is a parabolic subgroup of G, which is a proper one if $\dim X > 0$. If $\dim X = 0$, then $X = \{x\}$. Let $v \in V$ define x, and put $V_1 = V/kv$. Then G operates in V_1 and we can find a closed orbit in $\mathbb{P}(V_1)$. Continuing in this way one finds either a complete homogeneous space of dimension > 0, or G is isomorphic to a subgroup of a group of triangular matrices, hence is solvable. This implies the proposition.

7.2.3. Lemma. *Let P be a parabolic subgroup of G. Then G/P is a projective variety.*
By 5.2.3, G/P is quasi-projective, i.e. an open subvariety of a projective variety. The assertion now follows from 7.1.1(iv).

7.2.4. Lemma. *Let P and Q be closed subgroups of G, with $P \subset Q$. If P is parabolic in Q and Q is parabolic in G, then P is parabolic in G.*
We have to show that for any variety X, the projection $G/P \times X \to X$ is closed. Using 4.3.3(i) one sees that this is tantamount to proving that for any closed set $A \subset G \times X$ such that $(g,x) \in A$ implies $(gP,x) \subset A$, the projection A' of A onto X is closed.
Consider the morphism $\alpha: Q \times G \times X \to G \times X$ with $\alpha(q,g,x) = (gq,x)$. If A is as above, then $\alpha^{-1}A = \{(q,g,x) | (gq,x) \in A\}$, which is closed in $Q \times G \times X$. The completeness of Q/P implies

that the projection of $\alpha^{-1}A$ onto $G \times X$, i.e. the set $\bigcup_{(g,x)\in A} (gQ,x)$ is closed in $G \times X$. The completeness of G/Q implies that the projection on X of the last set is closed. Since this last projection coincides with A', the lemma is established.

7.2.5. _Theorem_ (_Borel's fixed point theorem_). _Let_ G _be_ _a_ _connected solvable linear algebraic group and_ X _a complete_ G-_space_. _There exists a point of_ X _which is fixed by all elements of_ G.

By 4.3.1(ii) there is closed orbit Y for the action of G on X. Let $x \in Y$, let P be the isotropy group of x in G. As in the proof of 7.2.2 one sees that P is a parabolic subgroup. By 7.2.2 it follows that $P = G$, whence $X = \{x\}$. This proves the theorem.

A _Borel subgroup_ of G is a connected, solvable, closed subgroup of G, which is maximal for these properties. A _maximal torus_ of G is a closed subtorus of G of maximal dimension. This notion occurred already in 6.11, in the case that G is connected and solvable.

7.2.6. _Theorem_. (i) _A closed subgroup of_ G _is parabolic if and only if it contains a Borel subgroup_;

(ii) _If_ B _is a Borel subgroup of_ G _then_ G/B _is complete_;

(iii) _All Borel subgroups of_ G _are conjugate_;

(iv) _All maximal tori of_ G _are conjugate_.

We may assume G to be connected. Let B be a Borel subgroup of G, and P any parabolic subgroup. Applying 7.2.5 to B and the complete variety G/P, we see that there is $x \in G$ such that $BxP = xP$. This means that $x^{-1}Bx \subset P$. Since $x^{-1}Bx$ is also a Borel subgroup, we have proved one half of (i). To prove the remaining part of (i) it suffices to prove (ii) (for if G/B is complete and $P \supset B$, then 7.1.1(iii), applied to the morphism $G/B \to G/P$, shows that G/P is complete).

We may assume that G is non-solvable. By 7.2.2 there exists a proper parabolic subgroup P of G, and replacing P by a conjugate, we may assume that $P \supset B$. Clearly, B is a Borel subgroup of P. By an induction on dim G we may assume that P/B is complete, i.e. that B is parabolic in P. It then follows from 7.2.4 that B is parabolic in G, establishing (ii).

If B and B' are two Borel subgroups, then (i) and (ii) imply that B' is conjugate to a subgroup of B and B' to a subgroup of B. Hence dim B = dim B', which implies (iii).

If T is a maximal torus, it is contained in a Borel subgroup (by the definition of Borel subgroup). Then (iv) follows from (iii) and the conjugacy of maximal tori of connected solvable groups (6.11(iii)).

7.2.7. *Corollary*. *Let* $\phi: G \to G'$ *be a surjective homomorphism of linear algebraic groups*. *Let* H *be a Borel subgroup (parabolic subgroup, maximal torus, respectively) of* G. *Then* ϕH *is a subgroup of* G' *of the same type*.

If H is a Borel subgroup of G then ϕH is closed, connected and solvable. Moreover the induced morphism $G/H \to G'/\phi H$ is surjective. By 7.1.1(iii), $G'/\phi H$ is complete, so 7.2.6(i) shows that ϕH contains a Borel subgroup. It follows that ϕH

is a Borel subgroup. It then also follows at that ϕH is parabolic if H is so.
Now let H be a maximal torus. Take a Borel subgroup B containing H. By 6.11(iv) we have a decomposition $B = H.B_u$, whence $\phi B = \phi H.\phi B_u$. Since ϕB is a Borel subgroup and ϕB_u is unipotent, we conclude from 6.11(iv) that ϕH must be a maximal torus ϕB, hence of H'.

Let Z(G) be the center of G, let B be a Borel subgroup of G.

7.2.8. Corollary. If G is connected then $Z(G)^0 \subset Z(B) \subset Z(G)$.
$Z(G)^0$ is connected, solvable, closed, hence contained in some Borel subgroup. But then it is contained in all Borel subgroups, by the conjugacy result 7.2.6(iii). This proves the first inclusion.
Let $g \in Z(B)$. The morphism $\phi: x \mapsto gxg^{-1}x^{-1}$ of G into G is such that $\phi(xB) = \phi(x)$, hence induces a morphism $G/B \to G$, which must be constant by 7.1.1. This means that $Z(B) \subset Z(G)$.

7.2.9. Corollary. If B is nilpotent, then $G^0 = B$.
We may assume G to be connected. We prove 7.2.9 by induction on dim G. If $B = \{e\}$ then $G = G/B$ is complete, whence $G = \{e\}$. If $B \neq \{e\}$ is nilpotent it follows from 6.5(ii) that $\dim Z(B) > 0$, whence $\dim Z(G) > 0$ by the preceding result. Replace G by $G/Z(G)^0$ and use induction.

Let T be a maximal torus of G. The connected centralizer $C = Z_G(T)^0$ of T in G is called a Cartan subgroup of G. It follows from 7.2.6(iv) that all Cartan subgroups of G are conjugate.

7.2.10. Corollary. C is nilpotent and equal to its connected normalizer $N_G(C)^0$.

T is a maximal torus of C and lies in the center of C. It follows (see 6.9(ii)) that the Borel subgroups of C are nilpotent. Hence C is nilpotent by the previous result. That $C = N_G(C)^0$ follows from the rigidity of tori (2.5.10).

7.2.11. Exercises. G is a connected linear algebraic group.

(1) Let H be a closed subgroup of G containing a maximal torus T. Then $N_G H \subset H^0 . N_G T$.

(2) Let B be a Borel subgroup of G. Then $(N_G B)^0 = B$.

(3) If all elements of G are semi-simple then G is a torus.

(4) Let $G = \mathbb{GL}_n$.

(a) The group $\mathbb{T}_n$ of upper triangular matrices in $\mathbb{GL}_n$ is a Borel subgroup of G;

(b) Let G act on $V = k^n$. A flag in V is a sequence of district subspaces V_i $(1 \leqslant i \leqslant s)$ of V such that $V_1 \subset V_2 \subset \ldots \subset V_s$. Then G also acts on the set of flags. The flag is complete if $s = n-1$ and $\dim V_i = i$. The Borel subgroups of G are the subgroups whose elements fix a given complete flag.

(Hint for the proof of (a): Let G act on $\mathbb{P}^{n-1}$. Use that the isotropy group of a point of $\mathbb{P}^{n-1}$ is a parabolic subgroup, and show by induction on n that $\mathbb{GL}_n / \mathbb{T}_n$ is complete).

(5) Assume char $k \neq 2$. Let $G = \mathbb{SO}_n$, viewed as a subgroup of $\mathbb{GL}_n$ fixing the symmetric bilinear form $\langle \, , \rangle$ of 6.6(1). A subspace W of $V = k^n$ is isotropic if the restriction of this form to W is zero. The Borel subgroups of G are the subgroups whose elements fix a given complete isotropic flag, i.e. a

maximal flag consisting of isotropic subspaces. (Hint: proceed as in the preceding exercise; one has to use Witt's theorem, see [26 , Ch. XIV, §5]).

(6) Establish s similar result for the symplectic group Sp_{2n} (see also 6.16(2)).

(7) Assume that G is defined over $\mathbb{F}_q$ and let F be the corresponding Frobenius morphism (see 3.3.15). There exists a maximal torus T of G and a Borel subgroup B ⊃ T such that FT = T, FB = B. If (T_1,B_1) is another such pair, there exists $x \in G$ with $Fx = x$ such that $T_1 = xTx^{-1}$, $B_1 = xBx^{-1}$ (Hint: Use Lang's theorem 3.3.16 and 7.2.6(iii),(iv)).

(8) Let B be a Borel subgroup of G and σ an automorphism of G (in the sense of algebraic groups) fixing all elements of B. Then σ = id.

(9) Let P and Q be parabolic subgroups of G, with P ⊂ Q. Assume that X is a closed subset of G such that XP = X. Then XQ is closed (Hint: use the completeness of G/P).

7.3. Further properties of Borel subgroups.

Let G be a connected linear algebraic group.

7.3.1. Lemma. *Let* H *be a closed subgroup of* G. *Put* $X = \bigcup_{x \in G} xHx^{-1}$.

(i) X *contains an open subset of its closure* $\overline{X}$. *If* G/H *is complete then* X *is closed*;

(ii) *If some element of* G *lies in only finitely many conjugates of* H *then* $\overline{X} = G$.

We may assume H to be connected. Then $Y = \{(x,y) \in G \times G \mid x^{-1}yx \in H\}$ is an irreducible closed subset of G × G, moreover if $(x,y) \in Y$ then $(xH,y) \subset Y$ (check these properties). Since the canonical morphism $G \times G \to G/H \times G$ is open (by 4.3.3(i)), it follows that the image Y_1 of Y in $G/H \times G$ i.e. the set $\{(xH,y) | x^{-1}yx \in H\}$ is closed in $G/H \times G$, and also irreducible. Let π: $G/H \times G \to G$ be the projection morphism. Since $X = \pi Y_1$, the assertions of (i) follow from 1.9.5 and the definition of completeness.

Since the fibers of the projection morphism $Y_1 \to G/H$ all have dimension dim H, it follows from 4.1.6(ii) that dim Y_1=dim G. Now let $x \in G$ have the property of (ii). Then $\pi^{-1}x$ is finite, and 4.2.5 shows that dim Y_1 = dim $\overline{X}$, so dim $\overline{X}$ = dim G and $\overline{X} = G$.

7.3.2. _Lemma_. _Let_ S _be_ _a_ _torus_ _in_ G. _There_ _exists_ $s \in S$ _such_ _that_ $Z_G(S) = Z_G S$.

It suffices to prove this for $G = GL(V)$. Then we have (by 2.5.2(c)) a direct sum decomposition $V = \bigoplus_{\chi \in A} V_\chi$, where A is a finite set of rational characters of S and

$$V_\chi = \{v \in V | s.v = \chi(s)v\}.$$

Now $Z_G S = \{g \in GL(V) | g.V_\chi = V_\chi\}$. Choose $s \in S$ such that $\chi(s) \neq \chi'(s)$ if $\chi, \chi' \in A$, $\chi \neq \chi'$ (check that this is possible). Then $Z_G(S) = Z_G S$.

7.3.3. _Theorem_. _Let_ G _be_ _a_ _connected_ _linear_ _algebraic_ _group_.

(i) _Every_ _element_ _of_ G _lies_ _in_ _a_ _Borel_ _subgroup_;

(ii) _Every_ _semi-simple_ _element_ _of_ G _lies_ _in_ _a_ _maximal_ _torus_;

(iii) _The_ _union_ _of_ _the_ _Cartan_ _subgroups_ _of_ G _contains_ _a_ _dense_ _open_ _subset_ _of_ G.

Let T be a maximal torus of G, $C = Z_G(T)^0$ the corresponding Cartan subgroup and $B \supset C$ a Borel subgroup (which exists, since C is nilpotent).
Take $t \in T$ such that $C = Z_G(t)^0$ (see the preceding lemma). If $x \in G$ and $t \in xCx^{-1}$, then t lies in the unique maximal torus xTx^{-1} of xCx^{-1}, hence $x^{-1}tx \in T$ and $C = (Z_G T)^0 \subset Z_G(x^{-1}tx)^0 = x^{-1}Cx$. This implies that $xCx^{-1} = C$, i.e. C is the only conjugate of C containing t. (iii) now follows from 7.3.1, applied to H = C. Since $\bigcup_{x\in G} xCx^{-1} \subset \bigcup_{x\in G} xBx^{-1}$, the latter set is dense in G, hence equal to G by 7.3.1(i), since G/B is complete. Finally, (ii) follows from (i) and 6.11(i).

7.3.4. _Corollary_. _Let_ B _be_ _a_ _Borel_ _subgroup_ _of_ G. _Then_ $Z(G) = Z(B)$.
If $x \in Z(G)$, then x lies in a Borel subgroup by 7.3.3(i), hence in all of them by the conjugacy of Borel subgroups. So $Z(G) \subset Z(B)$. The inclusion in the other direction was already proved in 7.2.8.

7.3.5. _Theorem_. _Let_ S _be_ _a_ _torus_ _in_ G.
(i) _The_ _centralizer_ $Z_G S$ _is_ _connected_;
(ii) _If_ B _is_ _a_ _Borel_ _subgroup_ _of_ G _containing_ S, _then_ $Z_G S \cap B$ _is_ _a_ _Borel_ _subgroup_ _of_ $Z_G S$, _and_ _all_ _Borel_ _subgroups_ _of_ $Z_G S$ _are_ _obtained_ _in_ _this_ _way_.
Take $x \in Z_G S$ and let B be a Borel subgroup containing x. Denote by X the fixed point set of x in G/B, i.e. $X=\{gB \mid xgB=gB\}$. Then $B \in X$, and X is a closed subvariety of G/B, which is complete (7.1.1(i)). Also, S acts on X and by 7.2.5 there is

a point of X fixed by S. This means that there is a Borel subgroup containing both x and S. It then follows from 6.10(ii) that x lies in $(Z_G S)^0$, whence (i).
Let B be as in (ii). Then $Z_G S \cap B$ is connected (6.12(ii)) and solvable. To prove the first part of (ii) it suffices to show that $Z_G S/Z_G(S) \cap B$ is complete (7.2.6). There is a bijective morphism of homogeneous spaces of $Z_G S/Z_G S \cap B$ onto the image of $Z_G S.B$ in G/B. Using 7.2.1 and the openness of the morphism $G \to G/B$ we see that it suffices to prove that $Y = Z_G S.B$ is closed in G. This set is irreducible (as the image of $Z_G S \times B$ under a morphism). If $y \in Y$, we have $y^{-1}Sy \subset B$. This also holds, by continuity, if $y \in \overline{Y}$. The morphism $\overline{Y} \times S \to B/B_u$ sending (y,s) to $y^{-1}syB_u$ satisfies the hypothesis of the rigidity theorem 2.5.10. It follows from that theorem that $y^{-1}sy \in sB_u$, if $y \in \overline{Y}$. This means that $y^{-1}Sy$ is a maximal torus of the algebraic group SB_u. By the conjugacy of maximal tori, there is a $z \in B_u$ such that $y^{-1}Sy = z^{-1}Sz$, from which we see that $y \in Z_G S.B$. Hence $\overline{Y} = Z_G S.B$, which proves the first assertion of (ii). The last one follows from the conjugacy of Borel subgroups of $Z_G S$.

Let T be a maximal torus.

7.3.6. <u>Corollary</u>. (i) $C = Z_G T$ <u>is a Cartan subgroup of</u> G; (ii) <u>If</u> B <u>is a Borel subgroup containing</u> T <u>then</u> $C \subset B$.
(i) and (ii) are consequences of 7.3.5(i), (ii), respectively.

7.3.7. <u>Theorem</u>. <u>Let</u> B <u>be a Borel subgroup of</u> G. <u>Then</u> $N_G B = B$.
Let $N = N_G B$. We prove that N = B by induction on dim G. Denote

by T a maximal torus of B. If $x \in N$, then xTx^{-1} is a maximal torus of B, hence there is a $b \in B$ with $xTx^{-1} = bTb^{-1}$. So, to prove that $x \in B$, we may assume that x normalizes T. Let ϕ be the homomorphism $T \to T$ with $\phi t = xtx^{-1}t^{-1}$. There are two cases: (a) ϕ is not surjective. Then $S = (\mathrm{Ker}\ \phi)^0$ is a nontrivial subtorus of T and $x \in Z_G S$. Clearly x normalizes the Borel subgroup $Z_G S \cap B$ of $Z_G S$. If $Z_G S \neq G$, we have $x \in B$ by induction. If $Z_G S = G$ then S lies in the center of G and passing to G/S we can again use induction.

(b) ϕ is surjective. Let $\rho: G \to GL(V)$ be a rational representation such that there is a $v \in V$ with

$$N = \{g \in G | \rho(g)v \in kv\}$$

(see 5.1.3). If $g \in B_u$ we have $\rho(g)v = v$ (since a unipotent group has no nontrivial rational characters). If $t \in T$ write $t = xt_1x^{-1}t_1^{-1}$, with $t_1 \in T$. One sees that also $\rho(t)v = v$. Hence $\rho(B)v = v$ and $gB \mapsto \rho(g)v$ defines a morphism $G/B \to V$. It follows from 7.1.1(iii),(vi) that this morphism is constant, hence $\rho(g)v = v$ for all v and $G = N$. But now 7.2.11(2) shows that $G = N = B$. This proves the theorem.

7.3.8. Corollary. _Let_ P _be a parabolic subgroup of_ G. _Then_ $P = N_G P$ _and_ P _is connected_.

Let $B \subset P^0$ be a Borel subgroup of G. If $x \in N_G P$, then xBx^{-1} is also a Borel subgroup of P^0, hence there is a $y \in P^0$ with $xBx^{-1} = yBy^{-1}$ (7.2.6(iii)). Then $y^{-1}x$ normalizes B, hence lies in B by the preceding theorem. So $x \in P^0$. This proves the corollary.

7.3.9. *Corollary*. *Let* P *and* Q *be parabolic subgroups of* G, *both containing a Borel subgroup* B. *If* P *and* Q *are conjugate in* G, *then* P = Q.

Put $Q = xPx^{-1}$. Then B and xBx^{-1} are two Borel subgroups of Q, hence $xBx^{-1} = yBy^{-1}$, for some $y \in Q$. As in the proof of the preceding corollary we find that $x \in Q$, whence P = Q.

7.3.10. Let $\mathcal{B}$ by the set of Borel subgroups of G and fix $B \in \mathcal{B}$. It follows from 7.2.6(iii) and 7.3.7 that the map $gB \mapsto gBg^{-1}$ defines a bijection of G/B onto $\mathcal{B}$. Via this bijection we can define a structure of projective variety on $\mathcal{B}$. It is readily seen that this is independent of the choice of B (check the details). We have thus defined the *variety of Borel subgroups* of G. It is a homogeneous space for G. More generally, fixing a parabolic subgroup $P \supset B$ we can define a structure of projective variety $\mathcal{P}$ on the set of conjugates of P. There is a surjective map $\mathcal{B} \to \mathcal{P}$, which is a morphism of homogeneous spaces ($\mathcal{P}$ being provided with the appropriate structure of homogeneous space).

7.3.11. *Exercises*. G is a connected linear algebraic group.

(1) Call $x \in G$ regular if the number of eigenvalues 1 of Ad(x) is as small as possible.

(a) The regular elements form a non-empty open subset of G.

(b) x is regular if and only if its semi-simple part x_s is regular.

(c) A semi-simple element is regular if and only if its centralizer has minimal dimension (here one has to use 4.4.4(ii)).

(d) A semi-simple element x is regular if and only if $Z_G(x)^0$ is a Cartan subgroup. (Hint: use 7.3.2 and 7.3.3(ii)).

(2) (a) A closed, nilpotent subgroup C of G such that $C = (N_G C)^0$ is a Cartan subgroup (Hint: use 6.16(2)).

(b) Let C be a maximal nilpotent subgroup of G such that each subgroup of finite index of C has finite index in its normalizer. Then C is a Cartan subgroup (This is a group-theoretical characterization of Cartan subgroups. Hint for the proof: show that the closure of a nilpotent subgroup of G is nilpotent, deduce that C is closed and then that C satisfies the conditions of (a)).

(3) Let $x = x_s x_u$ be an element of G. Show that $x_u \in Z_G(x_s)^0$, $x \in Z_G(x_s)^0$ (x_s and x_u are semi-simple and unipotent parts).

(4) Let char $k \neq 2$, $G = \mathbf{SO}_n$. Show that there exist semi-simple elements $x \in G$ such that $Z_G(x)$ is not connected (Hint: choose an appropriate element of order 2).

(5) Assume that T is a maximal torus of G, $C = Z_G T$ the corresponding Cartan subgroup and $B \supset C$ a Borel subgroup (see 7.3.6(ii)). Let $\sigma: G \to G$ be a surjective homomorphism of algebraic groups, such that $\sigma B = B$.

(a) Define a morphism $\phi_b: G \times B \to G$ by $\phi_b(x,c) = (\sigma x)b^{-1}cx^{-1}b$ ($b,c \in B$, $x \in G$). Then $\phi_b(e,e) = e$.

Show that the subspace $\operatorname{Im}(d\phi_b)_{(e,e)}$ of $\mathfrak{g}$ contains $\mathfrak{b}$ and $(\mathrm{Ad}(b)d\sigma-1)\mathfrak{g}$ (Hint: use 3.3.12).

(b) Let $T \subset B$ be a maximal torus. Show that there is $b \in B$ such that $\sigma T = b^{-1}Tb$ and that $(\mathrm{Ad}(b)d\sigma-1)^{-1}\mathfrak{b} \subset \mathfrak{b}$ (Hint: first take b such that the first condition is satisfied, modify it by an element of T and use that $C \subset B$). Deduce that for such a b the tangent maps $(d\phi_b)_{(e,e)}$ and $(d\phi_e)_{(e,b)}$ are surjective.

(c) Show that ϕ_e is surjective, i.e. that any element of G can be written in the form $(\sigma x)bx^{-1}$ ($x \in G, b \in B$) (Hint: adapt the argument used to prove 7.3.3(i), factor ϕ_e through $G/B \times G$).

(d) Using (c) prove that an arbitrary surjective homomorphism $\sigma: G \to G$ fixes a Borel subgroup.

Notes.

The elementary proof in 7.1.2 of the completeness of projective variety is taken from [20 , Ch.II, 5.6.1]. For a similar proof see [9].

The elegant result 7.1.4 is due to Chevalley. We have followed here the proof given in [30 , p.152-153].

In the proof of the basic properties of Borel subgroups contained in 7.2.6, due to Borel ([2]) we have avoided the use of flag varieties, which are needed in Borel's original proof (see [3 , p.261-262]). Instead, we use 7.2.2 and 7.2.4.

The proof of 7.3.7 is due to Borel (unpublished), it is also given in [23, p.144-145].

The results of exercise 7.3.11(5) are due to Steinberg [33 , no.47]. The proof sketched here is a bit different from the one of [loc. cit.].

8. <u>Linear action of a torus on a projective variety, applications</u>.

8.1. <u>Fixed points</u>.

Let ρ be a rational representation $\mathbb{G}_m \to GL(V)$. This representation defines an action of $\mathbb{G}_m$ on the projective space $\mathbb{P}(V)$, denoted by $(\xi,x) \mapsto \xi.x$. Let X be a closed subvariety of $\mathbb{P}(V)$, stable under the $\mathbb{G}_m$-action. We view $\mathbb{G}_m$ as the subset $\mathbb{P}^1-\{0,\infty\}$ of the projective line (see 1.6.12(2)).

8.1.1. <u>Lemma</u>. (i) <u>For any</u> $x \in X$ <u>the orbit map</u> $\xi \mapsto \xi.x$ <u>of</u> $\mathbb{G}_m$ <u>into</u> X <u>can be extended to a morphism</u> $\mathbb{P}^1 \to X$, <u>also denoted by</u> $\xi \mapsto \xi.x$;

(ii) $0.x$ <u>and</u> $\infty.x$ <u>are fixed points of</u> $\mathbb{G}_m$ <u>in</u> X. <u>Moreover</u>, $0.x = \infty.x$ <u>if and only if</u> x <u>is a fixed point</u>.

Choose a basis $(e_i)_{1\leqslant i\leqslant n}$ of V such that $\rho(\xi)e_i = \xi^{m_i}e_i$, with $m_i \in \mathbb{Z}$ such that $m_1 \geqslant m_2 \geqslant \dots \geqslant m_n$. Let $v = \sum_{i=a}^{b} \alpha_i e_i \in V$, with $\alpha_a \neq 0$, $\alpha_b \neq 0$. If $x = v^*$ is the point of $\mathbb{P}(V)$ defined by $v \in V$, we have

$$\xi.x = \Big(\sum_{i=a}^{b} \alpha_i \xi^{m_i-m_b} e_i\Big)^* = \Big(\sum_{i=a}^{b} \alpha_i (\xi^{-1})^{m_a-m_i} e_i\Big)^*$$

Put $f(\xi) = \big(\sum_{i=a}^{b} \alpha_i \xi^{m_i-m_b} e_i\big)^*, g(\xi) = \big(\sum_{i=a}^{b} \alpha_i \xi^{m_a-m_i} e_i\big)^*$. Then f and g define morphisms $\mathbb{A}^1 \to \mathbb{P}(V)$, such that $f(\xi) = g(\xi^{-1})$ if $\xi \neq 0$. They can be glued together to give a morphism $\phi\colon \mathbb{P}^1 \to \mathbb{P}(V)$, and $\phi(0) = \big(\sum_{m_i=m_b} \alpha_i e_i\big)^*$, $\phi(\infty) = \big(\sum_{m_i=m_a} \alpha_i e_i\big)^*$.

The morphism ϕ has the property required in (i).

It is clear that $\phi(0) = 0.x$ and $\phi(\infty) = \infty.x$ are fixed points.

Moreover, $0.x = \infty.x$ if and only if all m_i with $m_i \neq 0$ are equal, i.e. if and only if x is a fixed point. This proves (ii).

8.1.2. We shall prove below (8.1.4) a similar result for arbitrary tori. So let T be a torus and $\rho: T \to GL(V)$ a rational representation. This representation defines an action of T on $\mathbb{P}(V)$, which we write $(t,x) \mapsto t.x$. Let Y be a T-stable subvariety of $\mathbb{P}(V)$.
Denote by $X = X^*(T)$ the character group of T and by $X^\vee = X_*(T)$ the group of its 1-parameter multiplicative subgroups (see 2.5.1). X and $X^\vee$ are lattices (i.e. free abelian groups of finite ranks). They are in duality via the pairing $\langle\ ,\ \rangle$ defined by

$$\chi(\lambda(\xi)) = \xi^{\langle \xi,\lambda \rangle} \quad (\xi \in k\ ,\chi \in X,\lambda \in X^\vee)$$

(see 2.5.12(1),(2)).

8.1.3. *Lemma. There exist finitely many elements* $\chi_1,\ldots,\chi_r \in X$ *such that for all* $\lambda \in X^\vee$ *with* $\langle \chi_i,\lambda \rangle \neq 0$ $(1 \leq i \leq r)$ *the fixed points of* $\mathbb{G}_m$ *in* X*, acting via* $\rho \circ \lambda$*, coincide with the fixed points of* T.
If $\chi \in X$, let

$$V_\chi = \{v \in V | \rho(t)v = \chi(t)v,\ t \in T\},$$

then V is the direct sum of the nonzero V_χ (2.5.2(c)). It follows readily that the fixed points of T in X are the points of the form v^* with v in some V_χ. This implies that the 1-dimensional torus $\mathbb{G}_m$, acting via $\rho \circ \lambda$, will have the same fixed points as T if and only if $\langle \chi-\chi',\lambda\rangle \neq 0$ for all distinct χ,χ' which occur in V. This proves the lemma.

Let T and Y be as before.

8.1.4. <u>Proposition</u>. <u>Assume that</u> Y <u>is irreducible</u>, <u>and is not contained in any</u> $\mathbb{P}(W)$, W <u>a proper subspace of</u> V.
(i) <u>If</u> dim Y $\geq$ 1 <u>then</u> T <u>fixes at least</u> 2 <u>points of</u> Y;
(ii) <u>If</u> dim Y $\geq$ 2 <u>then</u> T <u>fixes at least</u> 3 <u>points of</u> Y.
By 8.1.3 we may assume that T = $\mathbb{G}_m$. It then follows from 8.1.1 that either T has at least two fixed points on Y, or that T fixes all of Y. This implies (i).
In the proof of (ii) we use the notations of the proof of 8.1.1. Let W be the subspace of V spanned by $e_2,\dots,e_n$. Since $\mathbb{P}(V) - \mathbb{P}(W)$ is isomorphic to $\mathbb{A}^{n-1}$ and Y is complete, we cannot have $Y \subset \mathbb{P}(V) - \mathbb{P}(W)$ if dim Y > 0, hence $Y \cap \mathbb{P}(W) \neq \phi$. All components of $Y \cap \mathbb{P}(W)$ are T-stable (because of the connectedness of T). If there is such a component Y' of dimension $\geq$ 1, then by (i) T has at least 2 fixed points in Y'. If $x \in Y-\mathbb{P}(W)$ then $\infty.x$ is a fixed point which is distinct from the previous ones. This establishes (ii), except for the case that $Y \cap \mathbb{P}(W)$ is finite. But then it follows from 4.2.5 (applied to a suitable affine open neighborhood of some $y \in Y \cap \mathbb{P}(W)$ and a morphism $U \to \mathbb{A}^1$ sending $(\Sigma\, \alpha_i e_i)^*$ to $\alpha_h^{-1}\alpha_1$, for some h) that dim Y = 1 (check the details). This finishes the proof.

Now let G be a connected linear algebraic group and T a maximal torus in G. The <u>Weyl group</u> of G relative to T is the finite group $W = N_GT/Z_GT$ (see 2.5.11). Let $\mathcal{B}$ be the variety of Borel subgroups of G (7.3.10), and denote by $\mathcal{B}_T$ the set of fixed points of T in $\mathcal{B}$, i.e. the set of Borel subgroups containing T. It is clear that N_GT operates on $\mathcal{B}_T$.

8.1.5. Lemma. (i) Z_GT *operates trivially on* $\mathcal{B}_T$;
(ii) *The action of* N_GT *on* $\mathcal{B}_T$ *defines an action of* W *on* $\mathcal{B}_T$ *which is simply transitive.*
(i) is a reformalation of 7.3.6(ii). If $xBx^{-1} \supset T$ then there is $b \in B$ such that $x^{-1}Tx = bTb^{-1}$ for some $b \in B$, so $n = xb \in N_GT$ and $xBx^{-1} = nBn^{-1}$. It follows that W acts transitivily on B_T. If $nBn^{-1} = B$, then (by 7.3.7) $n \in B \cap N_GT = N_BT = Z_BT$, which proves simple transitivity.

8.1.6. Proposition. (i) *Let* P *be a proper parabolic subgroup of* G. *Then* T *fixes at least* 2 *district points of* G/P *and at least* 3 *if* dim G/P > 1;
(ii) W = {e} *if and only* G *is solvable*;
(iii) *If* W *has order* 2 *then* dim $\mathcal{B}$ = 1;
(iv) G *is generated by the Borel subgroups containing* T.
Take a representation $\rho: G \to GL(V)$ such that P is the stabilizer of a line $L \subset V$, as in 5.1.3. Then X = G/P is the orbit of L in $\mathbb{P}(V)$. Now (i) follows from 8.1.4 and the preceding lemma. (iii) is a special case of (i). If G is solvable then W = {e} by 6.12(ii). If G is non-solvable then 7.2.2 and (i) imply that $W \neq \{e\}$. This proves (ii).
The subgroup Q of G generated by the Borel subgroups containing T is closed and connected (see 6.2), hence is para-

bolic. If $Q \neq G$ then (i) shows that there is a conjugate $xQx^{-1} \neq Q$, with $xQx^{-1} \supset T$. By an induction on dim G we may assume that xQx^{-1} is generated by Borel subgroups containing T, whence $xQx^{-1} = Q$, a contradiction. (iv) follows.

8.2. Groups of rank 1.

8.2.1. Let G be a linear algebraic group. The rank of G is the dimension of a maximal torus of G. The semi-simple rank of G is the rank of G/RG (whence RG is the radical, see 6.14). The results of 8.1 will be used to get detailed information about groups of rank 1. We shall, in particular, classify the semi-simple ones (8.2.4). We now assume that G is connected, of rank 1, and non-solvable. Let T be a maximal torus, and $B \supset T$ a Borel subgroup. The Weyl group $W = N_G T/Z_G T$ acts as a group of automorphisms on $T \simeq \mathbb{G}_m$, and since the only automorphisms of $\mathbb{G}_m$ are given by $\xi \mapsto \xi^{\pm 1}$ (check this) it follows that W has order ≤ 2. Then 8.1.6(ii) shows its order is 2. Take $n \in N_G T - Z_G T$. Then $ntn^{-1} = t^{-1}$ $(t \in T)$ and $n^2 = a \in Z_G T$. We denote by U the unipotent subgroup B_u of B. From 8.1.6(iii) we also see that dim G/B = 1.

8.2.2. Lemma. (i) G is the disjoint union of B and UnB;
(ii) $(U \cap nUn^{-1})^0$ is the radical of G;
(iii) $\dim U/U \cap nUn^{-1} = 1$.
Denote by $\xi \in G/B$ the point corresponding to B. From 8.1.5(ii) we see that $n\xi \neq \xi$, and that $\xi, n\xi$ are the two fixed points of T in G/B. It follows that $Un\xi \neq \{\xi\}$ and because

dim G/B = 1 we conclude from 1.9.5 that $Un\xi$ is the complement of a finite set $S \subset G/B$. Since $TUn\xi = Un\xi$, we have that T permutes the point of S, hence must fix them all. So $S \subset \{\xi, n\xi\}$, whence $Un\xi = G/B-\{\xi\}$. This implies (i). Since $U \cap nUn^{-1}$ is the isotropy group of ξ in U and since dim G/B=1, we have that $\dim U/U \cap nUn^{-1} = \dim U/(U \cap nUn^{-1})^0 = 1$. But then $(U \cap nUn^{-1})^0$ must be normal in U (as a consequence of 6.6(3)). This subgroup is also normalized by T and n, hence by B and nBn^{-1}. As these two groups generate G (by 8.1.6(iv)) we conclude that $(U \cap nUn^{-1})^0 \subset RG \subset B \cap nBn^{-1}$. Now (ii) follows by observing that RG cannot contain a torus. (iii) has already been established.

8.2.3. Lemma. *Assume, moreover, that G is semi-simple.*

(i) $Z_G T = T$, $U \cap nUn^{-1} = \{e\}$;

(ii) *There is a nonzero rational character* α *of* T *such that* $\mathfrak{g}$ *is the direct sum of* $\mathfrak{t}, \mathfrak{g}_\alpha$ *and* $\mathfrak{g}_{-\alpha}$, *where* $\mathfrak{g}_\beta = \{X \in \mathfrak{g} \mid Ad(t)X = \beta(t)X, t \in T\}$. *These subspaces are* 1-*dimensional and* $\mathfrak{g}_\alpha = \text{Lie } U$, $\mathfrak{g}_{-\alpha} = \text{Lie } nUn^{-1}$;

(iii) *The product map* $(u,b) \mapsto unb$ *is an isomorphism of varieties* $U \times B \to UnB = G-B$.

The previous lemma shows that dim U = 1 and that $U \cap nUn^{-1}$ is finite. Since the last group is normalized by T, it lies in the centralizer $Z_G T$. But $Z_G T$ is connected, contained in B and distinct from B (see 7.2.9), so we must have $Z_G T = T$. This implies (i).

We know by 2.6.6 that U is isomorphic to $\mathbb{G}_a$. Let $x: \mathbb{G}_a \to U$ be an isomorphism. There is a rational character α of T such that

$$tx(\xi)t^{-1} = x(\alpha(t)\xi) \quad (t \in T, \xi \in k),$$

and $\alpha \neq 1$ since $Z_G T = T$. If $X \in \mathfrak{g}$ is a nonzero element in $\mathrm{Im}(dx)$, then (check this)

$$\mathrm{Ad}(t)X = \alpha(t)X.$$

Clearly $X \in \mathrm{Lie}\ U$. Also, $\mathrm{Ad}(n)X \in \mathrm{Lie}\ nUn^{-1}$, and

$$(\mathrm{Ad}(t)\mathrm{Ad}(n)X = \alpha(t)^{-1}\mathrm{Ad}(n)X.$$

From 8.2.2(i) it follows that $\dim G \leq 3$. On the other hand, $\mathfrak{t} \oplus kX \oplus k\ \mathrm{Ad}(n)X$ is a 3-dimensional subspace of $\mathfrak{g}$. Since $\dim \mathfrak{g} = \dim G = 3$, the assertion of (ii) follows.

To prove (iii), it suffices to show that $(v,b) \mapsto vb$ is an isomorphism of $nUn^{-1} \times B$ onto $G\text{-}nB$. Put $V = nUn^{-1}$. We can view this morphism as an equivariant one of suitable homogeneous spaces for $V \times B$ (check this). It follows from 3.3.11 and (ii) that the tangent map at (e,e) is bijective. This implies (iii).

8.2.4. <u>Theorem. Let G be connected, semi-simple and of rank 1. Then G is isomorphic to $\mathbb{SL}_2$ or $\mathbb{PSL}_2$.</u>

We use the previous notations. Choose isomorphics $x: \mathbb{G}_a \xrightarrow{\sim} U$, $y: \mathbb{G}_m \xrightarrow{\sim} T$. Then $y(\eta)x(\xi)y(\eta)^{-1} = x(\eta^a\xi)$ $(\xi \in k, \eta \in k^*)$, so that $\alpha(y(\eta)) = \eta^a$. By 8.2.3(i) we have $a \neq 0$ and we may assume that $a > 0$. We also have $ny(\eta)n^{-1} = y(\eta^{-1})$. Putting $n^2 = y(\varepsilon)$ it follows that $y(\varepsilon) = ny(\varepsilon)n^{-1} = ny(\varepsilon^{-1})n^{-1}$, so $\varepsilon^2 = 1$.

It follows from 6.11(iv) and 8.2.3(iii) that $(\xi,\eta,\zeta) \mapsto x(\xi)ny(\eta)x(\zeta)$ defines an isomorphism of varieties $\mathbb{G}_a \times \mathbb{G}_m \times \mathbb{G}_a \tilde{\to} G-B$. If $\xi \neq 0$ then $nx(\xi)n^{-1} \notin B$, from which we conclude that there are rational functions $f,g,h \in k(T)$, defined for all $\xi \neq 0$, such that

$$(*) \quad nx(\xi)n^{-1} = x(f(\xi))ny(g(\xi))x(h(\xi)).$$

Conjugating both sides by $y(\eta)$ this implies

$$nx(\eta^{-a}\xi)n^{-1} = x(\eta^a f(\xi))ny(\eta^{-2}g(\xi))x(\eta^a h(\xi)),$$

from which one sees that $f(\eta^{-a}\xi) = \eta^a f(\xi), g(\eta^{-a}\xi) = \eta^{-2}g(\xi)$, $h(\eta^{-a}\xi) = \eta^a h(\xi)$. This shows that $\eta^{-2}g(1) = g(\eta^{-a})$. As $g(\xi) \neq 0$, this implies $a = 1$ or 2. Furthermore, $f(\xi) = \alpha\xi^{-1}$, $g(\xi) = \beta\xi^{-1}$, for some $\alpha,\beta \in k$. Taking inverses of both sides of $(*)$ it follows that $\alpha = \beta$ and $g(-\xi) = \varepsilon g(\xi)$. This implies that if $a = 1$ we have $g(\xi) = \gamma\xi^2$ and $\varepsilon = 1$, whereas if $a = 2$ we have $g(\xi) = \gamma\xi$ and $\varepsilon = -1$ (for suitable $\gamma \in k$). Replacing n by $ny(\lambda)$, with $\lambda \in k^*$ we may achieve that $\alpha = \beta = -1$. So

$$nx(\xi)n^{-1} = x(-\xi^{-1})ny(g(\xi))x(-\xi^{-1}) \quad (\xi \neq 0).$$

Let $\xi \neq 0,-1$ and apply this formula with $\xi,\xi+1$ and 1, respectively. Inserting the corresponding expressions in the formula $nx(\xi+1)n^{-1} = nx(\xi)n^{-1}.nx(1)n^{-1}$, we obtain after a straightforward computation that

$$-(\xi+1)^{-1} = -\xi^{-1} + g(\xi)^a(\xi^{-1}+1)^{-1},$$

whence $g(\xi)^a = \xi^2$.

(a) $a = 2$. We then have $g(\xi) = \gamma\xi$, with $\gamma^2 = 1$. Replacing n by $ny(\gamma)$ we are reduced to the case that $\gamma = 1$. Then

$$(**)\quad nx(\xi)n^{-1} = x(-\xi^{-1})ny(\xi)x(-\xi^{-1}),\ n^2 = y(-1).$$

Define a map ϕ: $G \to \mathbb{SL}_2$ by

$$\phi(x(\xi)y(\eta)) = \begin{pmatrix}1 & \xi\\0 & 1\end{pmatrix}\begin{pmatrix}\eta & 0\\0 & \eta^{-1}\end{pmatrix}$$

$$\phi(x(\xi)ny(\eta)x(\zeta)) = \begin{pmatrix}1 & \xi\\0 & 1\end{pmatrix}\begin{pmatrix}0 & 1\\-1 & 0\end{pmatrix}\begin{pmatrix}\eta & 0\\0 & \eta^{-1}\end{pmatrix}\begin{pmatrix}1 & \zeta\\0 & 1\end{pmatrix}.$$

The formula (**) is then the exact counterpart of the formula in $\mathbb{SL}_2$

$$\begin{pmatrix}0 & 1\\-1 & 0\end{pmatrix}\begin{pmatrix}1 & \xi\\0 & 1\end{pmatrix}\begin{pmatrix}0 & -1\\1 & 0\end{pmatrix} = \begin{pmatrix}1 & -\xi^{-1}\\0 & 1\end{pmatrix}\begin{pmatrix}0 & 1\\-1 & 0\end{pmatrix}\begin{pmatrix}\xi & 0\\0 & \xi^{-1}\end{pmatrix}\begin{pmatrix}1 & -\xi^{-1}\\0 & 1\end{pmatrix}.$$

The multiplication in the abstract group G is completely described by (**) and the other multiplication rules, given above (viz. $x(\xi+\eta) = x(\xi)x(\eta)$, $y(\xi\eta) = y(\xi)y(\eta)$, $y(\eta)x(\xi)y(\eta)^{-1} = x(\xi\eta^2)$, $ny(\eta)n^{-1}$, $n^2 = y(-1)$). Similarly for $\mathbb{SL}_2$.

It follows that ϕ defines an isomorphism of abstract groups $G \to \mathbb{SL}_2$, which is an isomorphism of varieties on the open set $G - B$.

It is clear that ϕ then also induces an isomorphism on any translate of G-B. Since these translates cover G, ϕ is an isomorphism of algebraic groups.

(b) $a = 1$. In this case we have $g(\xi) = \xi$ and $\varepsilon = 1$. So

$$nx(\xi)n^{-1} = x(-\xi^{-1})ny(\xi)x(-\xi^{-1}),\ n^2 = e.$$

Recall that $k[\mathbb{SL}_2] = k[T_1,T_2,T_3,T_4]/(T_1T_4-T_2T_3-1) = k[t_1,t_2,t_3,t_4]$ and that $k[\mathbb{PSL}_2]$ is the subalgebra generated by the products t_it_j $(1 \leq i,j \leq 4)$, see 2.1.4(3). It is more convenient to view $\mathbb{PSL}_2$ as the image of $\mathbb{SL}_2$ under the adjoint representation ρ. Its space is the Lie algebra $\mathfrak{sl}_2$, the space of 2×2-matrices with trace 0, and then $\rho(g)X = gXg^{-1}$ $(g \in \mathbb{SL}_2, X \in \mathfrak{sl}_2)$. On the basis $(e_i)_{1\leq i\leq 3}$ of $\mathfrak{sl}_2$ with $e_1 = \begin{pmatrix}1 & 0\\ 0 & -1\end{pmatrix}$, $e_2 = \begin{pmatrix}0 & 1\\ 0 & 0\end{pmatrix}$, $e_3 = \begin{pmatrix}0 & 0\\ 1 & 0\end{pmatrix}$, the linear transformation $\rho(g)$ is represented by the following 3×3-matrix $(a_{ij}(g))$:

$$\begin{pmatrix} x_1x_4+x_2x_3 & -x_1x_3 & x_2x_4 \\ -2x_1x_2 & x_1^2 & -x_2^2 \\ 2x_3x_4 & -x_3^2 & x_4^2 \end{pmatrix},$$

if $g = \begin{pmatrix}x_1 & x_2\\ x_3 & x_4\end{pmatrix}$. If $t_i \in k[\mathbb{SL}_2]$ is as above, we have $x_i = t_i(g)$. It readily follows that $k[\mathbb{PSL}_2]$ is also the subalgebra generated by the a_{ij}, which proves that $\rho(\mathbb{SL}_2)$ is isomorphic to $\mathbb{PSL}_2$.

Define a map $\phi\colon G \to \mathbb{PSL}_2$ by

$$\phi(x(\xi)y(\eta)) = \rho\left(\begin{pmatrix}1 & \xi\\ 0 & 1\end{pmatrix}\begin{pmatrix}\eta & 0\\ 0 & \eta^{-1}\end{pmatrix}\right)$$

$$\phi(x(\xi)ny(\eta)x(\zeta)) = \rho\left(\begin{pmatrix}1 & \xi\\ 0 & 1\end{pmatrix}\begin{pmatrix}0 & 1\\ -1 & 0\end{pmatrix}\begin{pmatrix}\eta & 0\\ 0 & \eta^{-1}\end{pmatrix}\begin{pmatrix}1 & \zeta\\ 0 & 1\end{pmatrix}\right)$$

As before one proves that ϕ is an isomorphism of algebraic groups. The details can be left to the reader.

8.2.5. Exercises. (1) Let G be as in 8.2.4. Show that G has

no proper normal closed subgroups of dimension > 0. It follows, in particular, that G equals its commutator subgroup and that $\mathbb{PSL}_2$ is simple (as an algebraic group).
(2) Let ρ be a non-trivial rational representation of $\mathbb{SL}_2$. Then Im ρ is isomorphic to $\mathbb{SL}_2$ or $\mathbb{PSL}_2$.
(3) Show that a 2-dimensional connected linear algebraic group is solvable.

8.2.6. We next deduce some consequences of 8.2.4, needed in the following chapters. We assume now that G is a connected reductive group, of semi-simple rank 1. The radical of G is a torus C (the identity component of the center), and G/C is semi-simple of rank 1. Moreover, $(G,G) \cap C$ is finite (see 6.15 for these facts).
Using 8.2.5(1) one sees that the canonical homomorphism π: $G \to G/C$ maps (G,G) surjectively onto G/C. Since $\operatorname{Ker} \pi \cap (G,G)$ is finite one concludes that (G,G) is connected, semi-simple and of rank 1, hence either $\mathbb{SL}_2$ or $\mathbb{PSL}_2$, by 8.2.4. Let T_1 be a maximal torus in (G,G) and $T \supset T_1$ one in G. Choose an isomorphism y: $G_m \xrightarrow{\sim} T_1$.
Using 8.2.3(ii) we see that there is a nontrivial rational character α of T such that, again, $\mathfrak{g} = \mathfrak{t} \oplus \mathfrak{g}_\alpha \oplus \mathfrak{g}_{-\alpha}$, where the subspaces $\mathfrak{g}_{\pm\alpha}$ are defined as before. They are 1-dimensional. We view y as an element of $X^\vee$. Then $\langle \alpha, y \rangle$ equals ± 1 or ± 2 (if G/C is of type $\mathbb{PSL}_2$ or $\mathbb{SL}_2$, respectively).
By 8.2.3(i) we have $Z_G T = T$. Choose $n \in N_{(G,G)}T_1 - T_1$, then $n^2 \in T_1$. Write $s(t) = ntn^{-1}$. Then s operates also on X and $X^\vee$, by

$$(s.x)(t) = x(s^{-1}.t), (s.x^\vee)(\xi) = s.x^\vee(\xi),$$

and we have $\langle s.x, s.x^\vee\rangle = \langle x, x^\vee\rangle$ $(x \in X, x^\vee \in X^\vee)$.

8.2.7. Lemma. (i) *There exists a unique* $\alpha^\vee \in X^\vee$ *such that*

$$s.x = x - \langle x,\alpha^\vee\rangle\alpha \ (x \in X).$$

We have $\langle\alpha,\alpha^\vee\rangle = 2$, *and* $s^2 = 1$;
(ii) n^2 *is independent of the choice of* n *and* $x(n^2) = (-1)^{\langle x,\alpha^\vee\rangle}$ $(x \in X)$;
(iii) *There exists a homomorphism of algebraic groups* x_α: $\mathbb{G}_a \to G$ *such that* $tx_a(\xi)t^{-1} = x_\alpha(\alpha(t)\xi)$ $(t \in T, \xi \in k)$ *and* $\mathrm{Im}(dx_\alpha) = \mathfrak{g}_\alpha$. *If* x'_α *has the same properties, there is a unique* $c \in k^*$ *such that* $x'_\alpha(\xi) = x_\alpha(c\xi)$;
(iv) T *and* $\mathrm{Im}\, x_\alpha$ *generate a Borel subgroup of* G, *whose Lie algebra is* $\mathfrak{t} \oplus \mathfrak{g}_\alpha$.
Let y be as above. Then

$$(s.x)(y(\xi)) = x(y(\xi))x((y(\xi))^{-2}) = x(y(\xi))\xi^{-2\langle x,y\rangle}.$$

If $\alpha^\vee$ is as in (i), the first member equals

$$x(y(\xi))\xi^{-\langle\alpha,y\rangle\langle x,\alpha^\vee\rangle} \ (x \in X, \xi \in k^*)$$

which shows that we must have

$$\langle\alpha,y\rangle\langle x,\alpha^\vee\rangle = 2\langle x,y\rangle.$$

If (G,G) is of type $\mathbb{SL}_2$, then $\langle\alpha,y\rangle = \pm 2$, and we must have $\alpha^\vee = \pm y$. If (G,G) is of type $\mathbb{PSL}_2$ we have $\langle\alpha,y\rangle = \pm 1$, and

$x^\vee = \pm 2y$. This proves uniqueness of $\alpha^\vee$.
For this choice of $\alpha^\vee$, we have

$$(s.x)(t) = x(t)\alpha(t)^{-\langle x,\alpha^\vee\rangle}$$

if $t \in T_1$ (by the definition of $\alpha^\vee$) and if $t \in C$ ($\alpha(C) = 1$ because of 4.4.7). Since $T = T_1C$ we obtain the formula of (i). The definition of $\alpha^\vee$ shows that $\langle\alpha,\alpha^\vee\rangle = 2$, and this implies $s^2 = 1$.
If n' is another choice of n, then $n' = nt$ for some $t \in T_1$, and $(n')^2 = n^2n^{-1}tnt = n^2$, since $ntn^{-1} = t^{-1}$. Put $X_1 = \{x \in X|\langle x,y\rangle = 0\}$, then X/X_1 is the character group of T_1 (check this). The formula for $x(n^2)$ is obvious for $x \in X_1$. It is also true if $C = \{e\}$, by a check in $\mathbb{SL}_2$ or $\mathbb{PSL}_2$. This implies (ii).
If x_α is as in (iii), then $tx_\alpha(\xi)t^{-1}x_\alpha(\xi)^{-1} = x_\alpha((\alpha(t)-1)\xi)$, which shows that $\mathrm{Im}\ x_\alpha \subset (G,G)$. The existence and uniqueness statements of (iii) therefore have only to be proved in $\mathbb{SL}_2$ or $\mathbb{PSL}_2$, and may be left to the reader. One uses that $T_1(\mathrm{Im}\ x_\alpha)$ is a Borel subgroup of (G,G). The last point (iv) also follows readily.
We terminate this section with auxiliary results about $\mathbb{SL}_2$. Let $G = \mathbb{SL}_2$. Then $B = \left\{\begin{pmatrix} x & y \\ 0 & x^{-1}\end{pmatrix} | x \in k^*, y \in k\right\}$ is a Borel subgroup. Its commutator subgroup is the unipotent group B_u, and the rational characters of B have the form $\begin{pmatrix} x & y \\ 0 & x^{-1}\end{pmatrix} \mapsto x^m$, for some $m \in \mathbb{Z}$ (check these facts).

8.2.8. <u>Lemma</u>. <u>Let</u> $f \in k[G]$ <u>be a nonzero regular function on</u> $G = \mathbb{SL}_2$ <u>such that there exists a rational character</u> χ <u>of</u> B

<u>with</u> $f(gb) = f(g)\chi(b)$ $(g \in G, b \in B)$. <u>Then</u> $\chi\begin{pmatrix} x & y \\ 0 & x^{-1} \end{pmatrix} = x^m$, <u>with</u> $m \geq 0$.

If $xt-yz = 1$, $z \neq 0$ then

$$\begin{pmatrix} x & y \\ z & t \end{pmatrix} = \begin{pmatrix} 1 & -z^{-1}x \\ 0 & 1 \end{pmatrix}\begin{pmatrix} 0 & -1 \\ 1 & 0 \end{pmatrix}\begin{pmatrix} z & t \\ 0 & z^{-1} \end{pmatrix},$$

and it follows that there exist $f \in k[T]$ and $m \in \mathbb{Z}$ with

$$f\begin{pmatrix} x & y \\ z & t \end{pmatrix} = z^m g(z^{-1}x).$$

In particular, $f\begin{pmatrix} 1 & 0 \\ z & 1 \end{pmatrix} = z^m g(z^{-1})$, if $z \neq 0$. If f is nonzero and regular on G, the last function must be defined for $z = 0$, which can only happen if $m \geq 0$. Notice that if f is non-constant we have $m > 0$.

8.2.9. <u>Exercises</u>. (1) Deduce from 8.2.8 a similar statement for $\mathbb{PSL}_2$.

(2)(a) In the situation of 8.2.8 show that the f with a prescribed value of $m \geq 0$ form a subspace V_m of $k[G]$ of dimension $m+1$, which is stable under left translations.

(b) Let ρ_m be the rational representation of G in V_m by left translations.

There exists a basis $(e_0, \ldots, e_m)$ of V_m such that

$$\rho_m\begin{pmatrix} x & 0 \\ 0 & x^{-1} \end{pmatrix} e_i = x^{m-2i} e_i \quad (x \in k^*),$$

$$\rho_m\begin{pmatrix} 1 & x \\ 0 & 1 \end{pmatrix} e_i = \sum_{j=0}^{i} (-1)^{i-j}\binom{i}{j} x^{i-j} e_j \quad (x \in k).$$

(c) Let $p = \text{char } k$. If $p = 0$ or $p > m$ then ρ_m is irreducible.

(3) Let (ρ, V) be a rational representation of $G = \mathbb{SL}_2$ and (ρ^*, V^*) the dual representation, i.e. V^* is the dual of the

vector space V and $(\rho^*(g)u)(v) = u(\rho(g)^{-1}v)$, if $v \in V$, $u \in V^*$, $g \in G$.

(a) Deduce from 6.7 that there exist $v \in V$, $v \neq 0$ and $m \in \mathbb{Z}$ such that $\rho\begin{pmatrix} x & y \\ 0 & x^{-1}\end{pmatrix}v = x^m v$. Define a linear map $\phi: V^* \to k[G]$ by $(\phi u)(g) = u(\rho(g)v)$. Show that $m \geqslant 0$.

(b) If ρ is irreducible then ϕ is injective.

(c) Any irreducible rational representation of G is equivalent to a quotient representation and also to a subrepresentation of some ρ_m.

(d) If char $k = 0$ every irreducible rational representation of G is equivalent to some ρ_m.

(4) Notations as in the previous exercises.

(a) V_m^* has a basis $(e_0^*,\ldots,e_m^*)$ with the following properties

$$\rho_m\begin{pmatrix} x & 0 \\ 0 & x^{-1}\end{pmatrix}e_i^* = x^{-m+2i}e_i^* \quad (x \in k^*)$$

$$\rho_m^*\begin{pmatrix} 1 & x \\ 0 & 1\end{pmatrix}e_0^* = \sum_{i=0}^{m} x^i e_i^* \quad (x \in k).$$

If $(\tilde{e}_i^*)$ is another basis with these properties, there exists $a \in k^*$ with $\tilde{e}_i^* = ae_i$.

(b) Let (ρ,V) be a rational representation of $G = \mathbb{SL}_2$ with the following properties: there is $v \in V, v \neq 0$ with $\rho\begin{pmatrix} x & 0 \\ 0 & x^{-1}\end{pmatrix}v = x^m v$, and $m \neq 0$, $\dim V = m+1$ and the vectors $(\rho(g)-1)v$ $(g \in G)$ span V. Then ρ is equivalent to ρ_m^*.

Notes.

The proof of the classification theorem 8.2.4 of semi-simple groups of rank 1 is different from the usual one. We exploit here "Bruhat's lemma" (8.2.2(i)). The argument of the proof of 8.2.4 goes back to Zassenhaus (see [19, p.393]).

It is quite convenient to have 8.2.4 available at an early stage. For example, as a consequence one gets the reflections of 8.2.7.

9. Roots, the Weyl group.

9.1. Roots.

9.1.1. Let G be a connected linear algebraic group, and T a maximal torus. X and $X^\vee$ are as before (8.1.2). We shall be concerned with subtori S of T of codimension 1, i.e. with dim S = dim T-1.

If S is any torus, acting linearly in a vector space V, the weights of S in V are the characters $\chi \in X^*S$ such that the space

$$V_\chi = \{v \in V | s.v = \chi(s)v\}$$

is nonzero. A nonzero $v \in V_\chi$ is called a weight vector of S, with weight χ. By 2.5.2(c), V is the direct sum of the V_χ (χ running through the weights).

9.1.2. Lemma. G is generated by the centralizers $Z_G S$, where S runs through the codimension 1 subtori of T.

Recall that the $Z_G S$ are connected (7.3.5(i)). Let $\chi_1, \dots, \chi_r$ be the distinct nonzero weights of T in the Lie algebra $\mathfrak{g}$. Then $\mathfrak{g}$ is the direct sum of the spaces $\mathfrak{g}_{\chi_i}$ $(1 \leq i \leq r)$ and $\mathfrak{g}_0$ (the fixed point set of Ad(T)). Let $S_i = (\mathrm{Ker}\ \chi_i)^0$, this is a subtorus of T of codimension 1. It follows from 4.4.7 that $\mathrm{Lie}(Z_G S_i)$ contains $\mathfrak{g}_0$ and $\mathfrak{g}_{\chi_i}$. The Lie algebra of the subgroup G_1 generated by the $Z_G S_i$, which is closed and connected (by 6.1), coincides with $\mathfrak{g}$, hence $G_1 = G$, proving the lemma.

A codimension 1 subtorus of T is regular if its centralizer is solvable and singular otherwise.

9.1.3. Lemma. *Let $S \subset T$ be a singular subtorus of codimension 1. There is a nonzero weight α of T in $\mathfrak{g}$ such that $S=(\mathrm{Ker}\ \alpha)^0$ and that $\mathrm{Lie}(Z_G S/R_u(Z_G S)) \cong \mathrm{Lie}(Z_G T) \oplus kX_\alpha \oplus kX_{-\alpha}$, where X_α and $X_{-\alpha}$ are weight vectors for α and $-\alpha$, respectively. α is unique up to sign.*

It is immediate that $Z_G S$ is of semi-simple rank 1. So $Z_G S/R_u(Z_G S)$ is reductive, of semi-simple rank 1, and we have the asserted decomposition of its Lie algebra by 8.2.6. It also follows that $S \subset \mathrm{Ker}\ \alpha$. Since codim $S = 1$, we must have $S = (\mathrm{Ker}\ \alpha)^0$. These remarks imply the lemma.

9.1.4. The characters $\alpha \in X$ obtained, in the manner of 9.1.3, from singular tori of codimension 1 are called the *roots* of G with respect to T. Let $R = R(G,T)$ be the set of roots. It is a finite subset of X: from the fact that the homomorphism $Z_G S \to Z_G S/R_u(Z_G S)$ (in the situation of 9.1.3) is separable it follows that the roots occur among the weights of T in $\mathrm{Lie}(Z_G S)$ and these occur among the finite set of weights of T in $\mathfrak{g}$ (check this argument). Denote by $W = W(G,T)$ the Weyl group $N_G T/Z_G T$ (see 8.1). It acts on T, and also on $X, X^\vee$ by

$$(w.x)(t) = x(w^{-1}.t),\ (w.x^\vee)(\xi) = w.x^\vee(\xi),$$

and $\langle w.x, w.x^\vee\rangle = \langle x, x^\vee\rangle \quad (x \in X, x^\vee \in X^\vee, \xi \in k^*)$.

It is clear from the definition that W permutes the elements of R. Now let $\alpha \in R$ and put $G_\alpha = Z_G(\mathrm{Ker}\ \alpha)^0$. Since G_α is non-solvable, $N_{G_\alpha} T \neq Z_{G_\alpha} T$. Choose $n_\alpha \in N_{G_\alpha} T - Z_{G_\alpha} T$ and let s_α be the element of W defined by n_α.

9.1.5. Lemma. (i) *There is a unique $\alpha^\vee \in X^\vee$ such that*

$\langle \alpha, \alpha^\vee \rangle = 2$ *and that*

$$s_\alpha . x = x - \langle x, \alpha^\vee \rangle \alpha, \quad s_\alpha . x^\vee = x^\vee - \langle \alpha, x^\vee \rangle \alpha^\vee \quad (x \in X, x^\vee \in X^\vee).$$

We have $s_{-\alpha} = s_\alpha$ *and* $s_\alpha^2 = 1$;

(ii) *If* $w \in W$ *then* $w s_\alpha w^{-1} = s_{w.\alpha}$, $w.\alpha^\vee = (w.\alpha)^\vee$.

The existence of $\alpha^\vee$ and the first formula of (i) follow from 8.2.7(i), applied to the group $G_\alpha / R_u G_\alpha$. The second formula is a consequence of $\langle s_\alpha . x, s_\alpha . x^\vee \rangle = \langle x, x^\vee \rangle$. That $s_{-\alpha} = s_\alpha$ follows from the definition of n_α, and it is trivial to check from the formulas that $s_\alpha^2 = 1$ (one can also use that $n_\alpha^2 \in Z_{G_\alpha} T$). Let $n \in N_G T$ represent $w \in W$. Then $G_{w.\alpha} = n G_\alpha n^{-1}$, and this implies that $s_{w.\alpha} = w s_\alpha w^{-1}$. The second formula of (ii) is a consequence of this.

The $\alpha^\vee \in X^\vee$ (for $\alpha \in R(G,T)$ are called the *coroots* of G with respect to T.

9.1.6. A *root datum* is a quadruple $\Psi = (X, R, X^\vee, R^\vee)$ where X and $X^\vee$ are free abelian groups of finite rank, in duality by a pairing $X \times X^\vee \to \mathbb{Z}$ denoted by $\langle \, , \rangle$, R and $R^\vee$ are finite subsets of X and $X^\vee$. Moreover we assume given a bijection $\alpha \mapsto \alpha^\vee$ of R onto $R^\vee$.

If $\alpha \in R$ define endomorphisms s_α of X and $X^\vee$ by

$$s_\alpha x = x - \langle x, \alpha^\vee \rangle \alpha, \quad s_\alpha x^\vee = x^\vee - \langle \alpha, x^\vee \rangle \alpha^\vee.$$

The following two axioms are imposed:

(RD1) *If* $\alpha \in R$ *then* $\langle \alpha, \alpha^\vee \rangle = 2$; $\Rightarrow 0 \notin R, 0 \notin R^\vee$

(RD2) *If* $\alpha \in R$ *then* $s_\alpha R \subset R$, $s_\alpha R^\vee \subset R^\vee$.

It is clear form (RD1) that $s_\alpha^2 = 1$, $s_\alpha \alpha = -\alpha$.

Let Ψ be as above. Denote by Q the subgroup of X generated by R and put $V = Q \otimes_{\mathbb{Z}} \mathbb{R}$. If $R \neq \phi$ then R is a root system in V in the sense of [8,Ch.VI,no.1]. This means that the following axioms are satisfied:

(RS1) R is finite and generates V, moreover $0 \notin R$;

(RS2) If $\alpha \in R$ there is $\alpha^{\vee} \in V^{\vee}$ (the dual of V) such that $\langle \alpha,\alpha^{\vee} \rangle = 2$ and that s_{α} (defined as before) stabilizes R;

(RS3) If $\alpha \in R$ then $\alpha^{\vee}(R) \subset \mathbb{Z}$.

The rank of R is by definition the dimension of V.

Let $X, X^{\vee}$, R be as in 9.1.4. It follows from 9.1.5(i) that there is a subset $R^{\vee} \subset X^{\vee}$ such that (RD1) and (RD2) hold. Moreover, in this case the map $\alpha \mapsto \alpha^{\vee}$ is bijective. In fact, if $\alpha^{\vee} = \beta^{\vee}$ then $s_{\alpha}s_{\beta}x = x + \langle x,\alpha^{\vee} \rangle(\alpha-\beta)$. Since $\langle \alpha-\beta,\alpha^{\vee} \rangle = 0$, it follows that $s_{\alpha}s_{\beta}$ is a unipotent transformation (in $X \otimes \mathbb{C}$). As the Weyl group W is finite, we must have $s_{\alpha}s_{\beta} = \mathrm{id}$, whence $\alpha = \beta$.

So we have associated to G and T a root datum $\Psi = \Psi(G,T)$. From the conjugacy of maximal tori one infers that Ψ is uniquely determined by G, up to isomorphism. The corresponding root system is called the root system R(G,T) of R with respect to T. It has the extra property of being reduced, i.e. it has the property of the following lemma.

9.1.7. Lemma. *Let R be as in 9.1.4. If $\alpha \in R$ and $r\alpha \in R$ for some $r \in \mathbb{Q}$ then $r = \pm 1$.*

Since $(\mathrm{Ker}\ \alpha)^0 = (\mathrm{Ker}\ r\alpha)^0$ we have, by the definition of roots, that weights α and $r\alpha$ occur in $\mathrm{Lie}(Z_G(\mathrm{Ker}\ \alpha)^0/R_uZ_G(\mathrm{Ker}\ \alpha)^0)$. According to 9.1.3 this can only be if $r = \pm 1$.

.1.8. Let $\Psi = \Psi(G,T)$ be as before. The Weyl group $W = W(G,T)$ operates on the ingredients of Ψ. To study W, it is convenient to introduce a positive definite symmetric bilinear form (,) on $X_{\mathbb{R}} = X \otimes_{\mathbb{Z}} \mathbb{R}$, which is W-invariant. Such forms exist: let f be any positive definite symmetric bilinear form on $X_{\mathbb{R}}$ and take the form (,) defined by

$$(x,y) = \sum_{w \in W} f(w.x,w.y).$$

We thus have a Euclidean metric on $X_{\mathbb{R}}$ which is W-invariant, and the s_α ($\alpha \in R$) are Euclidean reflections. It follows (check this) that

$$\langle x,\alpha^\vee \rangle = 2(\alpha,\alpha)^{-1}(x,\alpha) \quad (x \in X_{\mathbb{R}}, \alpha \in R).$$

Consequently we could identify $R^\vee$ with the set $\{2(\alpha,\alpha)^{-1}\alpha \mid \alpha \in R\}$, and $\alpha^\vee$ with $2(\alpha,\alpha)^{-1}\alpha$. We can now establish the following theorem.

9.1.9. Theorem. W(G,T) is generated by the s_α, $\alpha \in R(G,T)$. If W is trivial, G is solvable (8.1.6(ii)). The same is true if $R(G,T) = \emptyset$, by 9.1.2 and 7.3.5(ii). In these cases the theorem is trivially true. So assume W nontrivial and $R(G,T) \neq \emptyset$. We proceed by induction on dim G. Assume $w \in W$, $w \neq 1$ and let $n \in N_G T$ represent w. Consider the homomorphism $\phi: t \mapsto ntn^{-1}t^{-1}$ of T into itself. There are 2 cases:
(a) ϕ is not surjective. Then n centralizes a subtorus $S \neq \{e\}$ of T. If S lies in the center of G we can pass to G/S and apply induction. Otherwise, n lies in $Z_G S \neq G$. Since $R(Z_G S,T) \subset R(G,T)$, $W(Z_G S,T) \subset W(G,T)$ we may, by induction,

assume that w is a product of reflections.

(b) ϕ is surjective. This implies that the linear map w-1 of $X_{\mathbb{R}}$ is injective, hence is bijective. Take $\alpha \in R(G,T)$. There is $x \in X_{\mathbb{R}}$ with $(w-1)x = \alpha$. If (,) is as before, then

$$(x,x) = (wx,wx) = (x+\alpha,x+\alpha) = (x,x) + 2(x,\alpha) + (\alpha,\alpha),$$

and it follows that $\langle x,\alpha^\vee\rangle = -1$. Hence $s_\alpha x = x+\alpha = wx$, consequently $s_\alpha w$ fixes x. So this element of W is in case (a), which we have already dealt with. This proves the theorem.

9.1.10. Exercises. (1) Let $\Psi = (X,R,X^\vee,R^\vee)$ be a root datum. Define a homomorphism $p: X \to X^\vee$ by $px = \sum_{\alpha\in R} \langle x,\alpha^\vee\rangle\alpha^\vee$.

(a) Deduce from the identity

$$\langle \alpha,\beta^\vee\rangle^2\alpha^\vee = \langle \alpha,\beta^\vee\rangle\beta^\vee + \langle \alpha,s_\alpha(\beta^\vee)\rangle s_\alpha(\beta^\vee)$$

that $p\alpha = \frac{1}{2}\langle \alpha,p\alpha\rangle\alpha^\vee$ $(\alpha \in R)$. Show that $X_0 = \operatorname{Ker} p = \{x \mid \langle x,\alpha^\vee\rangle = 0,\ \alpha \in R\}$.

(b) Show that $Q \cap X_0 = \{0\}$ and that $Q+X_0$ has finite index in X.

(c) The Weyl group $W(\Psi)$ is the group of endomorphisms of X generated by the s_α. Show that it is a finite group.

Let G and T be as in 9.1.1, let $W = W(G,T)$.

(2) If $w \in W$ fixes $x \in X_{\mathbb{R}}$, it is a product of reflections s_α fixing x.

(3) An element $w \in W$ is a product of at most r reflections s_α, where r is the semi-simple rank of G (see 8.2.1).

9.2. Relative position of two roots.

In this section we deduce some information about the "relative position" of two roots, which we need later on. We assume that $\Psi = (X,R,X^{\vee},R^{\vee})$ is a root datum. The group W generated by the reflections s_{α} ($\alpha \in R$) is finite (see 9.1.10(1) compare also 9.1.9), let (,) be a positive definite symmetric bilinear form on $X_{\mathbb{R}}$ which is W-invariant.

9.2.1. Lemma. Let $\alpha,\beta \in R$ be two linearly independent roots.
(i) $0 \leq \langle \alpha,\beta^{\vee}\rangle\langle \beta,\alpha^{\vee}\rangle \leq 3$, hence $|\langle \alpha,\beta^{\vee}\rangle|$ equals 0,1,2,3. If $|\langle \alpha,\beta^{\vee}\rangle| > 1$, then $|\langle \beta,\alpha^{\vee}\rangle| = 1$;
(ii) $\langle \alpha,\beta^{\vee}\rangle = 0$ if and only if $\langle \beta,\alpha^{\vee}\rangle = 0$.

We have $s_{\alpha}s_{\beta}\alpha = (\langle \alpha,\beta^{\vee}\rangle\langle \beta,\alpha^{\vee}\rangle-1)\alpha - \langle \alpha,\beta^{\vee}\rangle\beta$, $s_{\alpha}s_{\beta}\beta = -\beta + \langle \beta,\alpha^{\vee}\rangle\alpha$, so $s_{\alpha}s_{\beta}$ stabilizes the subspace of $X_{\mathbb{R}}$ spanned by α and β. Its restriction to that subspace is represented on the basis (α,β) by the matrix

$$M_{\alpha\beta} = \begin{pmatrix} \langle \alpha,\beta^{\vee}\rangle\langle \beta,\alpha^{\vee}\rangle-1 & \langle \beta,\alpha^{\vee}\rangle \\ -\langle \alpha,\beta^{\vee}\rangle & -1 \end{pmatrix}.$$

Since $s_{\alpha}s_{\beta}$ has finite order, the eigenvalues of $M_{\alpha\beta}$ are two complex conjugate roots of unity, hence the absolute value of the trace of $M_{\alpha\beta}$ is at most 2. As $s_{\alpha} \neq s_{\beta}$, the eigenvalues cannot both be 1. It follows that

$$-2 \leq \langle \alpha,\beta^{\vee}\rangle\langle \beta,\alpha^{\vee}\rangle-2 < 2,$$

whence (i). If $\langle \alpha,\beta^{\vee}\rangle = 0$ then $M_{\alpha\beta}$ is triangular and can only be of finite order if also $\langle \beta,\alpha^{\vee}\rangle = 0$.

9.2.2. A subset R^+ of R is a <u>system of positive roots</u> in R if it has the following properties: (a) no positive linear combination $\sum_i n_i\alpha_i$, with $n_i > 0, \alpha_i \in R^+$, equals 0;
(b) $R = R^+ \cup (-R^+)$. Such subsets exist: let $x \in X_{\mathbb{R}}$ be such that $(x,\alpha) \neq 0$ for all $\alpha \in R$ and put $R^+ = \{\alpha \in R \mid (x,\alpha) > 0\}$. We shall see in 10.1.6 that any system of positive roots is of this kind. Suppose a set of positive roots R^+ has been fixed, and write $\alpha > 0$ if $\alpha \in R^+$ and $\alpha < 0$ otherwise.

9.2.3. <u>Lemma</u>. <u>Let</u> $\alpha,\beta \in R$ <u>be linearly independent</u>. <u>There is</u> $w \in W$ <u>such that</u> $w\alpha > 0$, $w\beta > 0$.
We may reduce things to the case that X is spanned by α and β. We put $a = \langle \alpha,\beta^\vee \rangle\langle \beta,\alpha^\vee \rangle$, then $a = 0,1,2,3$ by the previous lemma.
If $a = 0$, then by 9.2.1(ii) we have $\langle \alpha,\beta^\vee \rangle = \langle \beta,\alpha^\vee \rangle = 0$ and $-s_\alpha\alpha = s_\beta\alpha = \alpha$, $s_\alpha\beta = -s_\beta\beta = \beta$, from which the assertion readily follows.
Now assume $a \neq 0$. We may assume $\alpha < 0$. Then $s_\alpha\alpha = -\alpha > 0$, $s_\alpha\beta = \beta - \langle \beta,\alpha^\vee \rangle\alpha$. If $s_\alpha\beta > 0$ we are through. Otherwise, replacing α,β by $s_\alpha\beta, -\alpha$, respectively, we may assume we are in the case $\alpha < 0$, $\beta > 0$. We may also assume that $s_\alpha\beta < 0$. This implies, in particular, that $\langle \beta,\alpha^\vee \rangle < 0$.
Now $s_\alpha s_\beta\alpha = (a-1)\alpha - \langle \alpha,\beta^\vee \rangle\beta$, $s_\alpha s_\beta\beta = -\beta + \langle \beta,\alpha^\vee \rangle\alpha > 0$. If $a = 1$ then $s_\alpha s_\beta\alpha = -\langle \alpha,\beta^\vee \rangle\beta > 0$, and $w = s_\alpha s_\beta$ is as required.
If $a = 2,3$ then $(s_\alpha s_\beta)^a = -1$ (look at the matrix $M_{\alpha\beta}$ of the proof of 9.2.1). We may then, replacing if necessary α,β by $-\alpha,-\beta$, respectively, assume that $\langle \alpha,\beta^\vee \rangle = -1$, $\langle \beta,\alpha^\vee \rangle = -a$.
If $a = 2$ we have

$$s_\alpha\alpha = -\alpha \quad , s_\beta\alpha = \alpha+\beta, \ s_\alpha s_\beta\alpha = \alpha+\beta \quad ,$$
$$s_\alpha\beta = 2\alpha+\beta, \ s_\beta\beta = -\beta \ , \ s_\alpha s_\beta\beta = -2\alpha-\beta.$$

If $2\alpha+\beta > 0$ we took $w = s_\alpha$, if $\alpha+\beta < 0$ we may take $w = -s_\beta$ and if $2\alpha+\beta < 0$, $\alpha+\beta > 0$ we take $w = s_\alpha s_\beta$.

If $a = 3$ we have

$$s_\alpha\alpha=-\alpha \ , \ s_\beta\alpha=\alpha+\beta, \ s_\alpha s_\beta\alpha=2\alpha+\beta \ , \ s_\beta s_\alpha\alpha=-\alpha-\beta \ , \ s_\alpha s_\beta s_\alpha\alpha=-2\alpha-\beta$$
$$s_\alpha\beta=3\alpha+\beta, \ s_\beta\beta=-\beta \ , \ s_\alpha s_\beta\beta=-3\alpha-\beta, \ s_\beta s_\alpha\beta=3\alpha+2\beta, \ s_\alpha s_\beta s_\alpha\beta=3\alpha+2\beta.$$

If $3\alpha+\beta > 0$ we took $w = s_\alpha$, if $\alpha+\beta < 0$ we may take $w = -s_\beta$, if $3\alpha+\beta < 0$, $2\alpha+\beta > 0$ take $w = s_\alpha s_\beta$, if $\alpha+\beta > 0$, $3\alpha + 2\beta < 0$ take $w = -s_\beta s_\alpha$ and if $2\alpha+\beta < 0$, $3\alpha+2\beta > 0$ take $w = s_\alpha s_\beta s_\alpha$. This proves 9.2.3.

9.2.4. Exercise. Let α,β and a be as in 9.2.3. If $a > 0$, assume $\langle \alpha,\beta^\vee \rangle = -1$, $\langle \beta,\alpha^\vee \rangle = -a$. The set S described below is a subset of R which is stable under s_α and s_β.

$a = 0$, $S = \{\pm\alpha,\pm\beta\}$,

$a = 1$, $S = \{\pm\alpha,\pm\beta,\pm(\alpha+\beta)\}$.

$a = 2$, $S = \{\pm\alpha,\pm\beta,\pm(\alpha+\beta),\pm(2\alpha+\beta)\}$,

$a = 3$, $S = \{\pm\alpha,\pm\beta,\pm(\alpha+\beta),\pm(2\alpha+\beta),\pm(3\alpha+\beta),\pm(3\alpha+2\beta)\}$.

9.3. Borel groups and systems of positive roots, the unipotent radical.

9.3.1. G and T are as in 9.1.1. Let $B \supset T$ be a Borel subgroup of G. If $S \subset T$ is a singular codimension 1 subtorus of T then $B \cap Z_G S$ is a Borel subgroup of $Z_G S$ (7.3.5(ii)). It follows from 9.1.3 and 8.2.7(iv) that there is a root $\alpha \in R(G,T)$ such that $\mathrm{Lie}(B \cap Z_G S/R_u(Z_G S)) \cong t \oplus kX_\alpha$. Let $R^+(B)$ be the set of roots α obtained in this manner from the singular tori S.

9.3.2. Proposition. $R^+(B)$ is a system of positive roots in $R(G,T)$.

Let α be a root, let $G_\alpha = Z_G(\mathrm{Ker}\,\alpha)^0$. Since $B \cap G_\alpha$ is a Borel subgroup of G_α containing T, it follows that either α or $-\alpha$ lies in $R^+(B)$. Now let $\rho: G \to GL(V)$ be a rational representation such that there is $v \in V$ with $B = \{g \in G | \rho(b)v \in kv\}$ (see 5.1.3). Fix a basis $(e_1,\ldots,e_n)$ of V with $e_1 = v$, write $\rho(g)e_i = \sum_{j=1}^{n} a_{ij}(g)e_j$ and put $F(g) = a_{11}(g)$. Then $F \in k[G]$, $F(e) = 1$ and $F(gb) = \chi(b)F(g)$ $(g \in G, b \in B)$, where χ is a rational character of B. If $G \neq B$ (which, we may assume) we have $\chi \neq 1$ (check this, using the completeness of G/B). The restriction of χ to T defines a nonzero element $x \in X$.
Let $\alpha \in R^+(B)$ and define G_α as before. Since $F(gu) = F(g)$ if $g \in G_\alpha$, $u \in R_u G_\alpha$, it follows that F defines a regular function on $G_\alpha/R_u G_\alpha$ and also on its commutator subgroup H, which is isomorphic to $\mathbb{SL}_2$ or $\mathbb{PSL}_2$ (8.2.4). We have $\mathrm{Lie}\, G_\alpha/R_u G_\alpha = t \oplus kX_\alpha \oplus kX_{-\alpha}$, and Lie H contains X. Let $\alpha^\vee \in X^\vee$ be the coroot corresponding to α. By the definition of F we have

$$F(g\alpha^\vee(\xi)) = \xi^{\langle x,\alpha^\vee \rangle} F(g) \quad (g \in H, \xi \in k^*).$$

But now it follows from 8.2.8 and 8.2.9(1) that $\langle x,\alpha^\vee \rangle > 0$. Consequently, $(x,\alpha) > 0$ if $\alpha \in R^+(B)$. Since we have already seen that for any root $\beta \in R(G,T)$, either β or $-\beta$ lies in $R^+(B)$, it follows that $R^+(B) = \{\alpha \in R(G,T) | (x,\alpha) > 0\}$, which proves 9.3.2.

Take a root α and denote by H_α the identity component of the intersection of the groups B_u, where B runs through the Borel subgroups containing T with $\alpha \in R^+(B)$. Let V be the identity component of the intersection of all B_u $(B \supset T)$.

9.3.3. Lemma. $H_\alpha \supset V$ and V is a closed normal subgroup of H_α.
That $H_\alpha \supset V$ is obvious. Since T normalizes H_α, it follows that Lie H_α is spanned by elements of Lie V and weight vectors X_β, where β is a root. Such a β must have the property that $\beta \in R^+(B)$ for all Borel subgroups B with $\alpha \in R^+(B)$. Since the Weyl group W permutes transitively the Borel subgroups containing T, it follows that there is a system of positive roots, relative to which we have $w\beta > 0$ for all $w \in W$ with $w\alpha > 0$. By 9.2.3 and 9.1.7 we have $\alpha = \beta$. This implies that $\dim H_\alpha/V \leq 1$, and consequently V is normal in H_α (6.6(3)).

We can now prove the fundamental theorem about the unipotent radical of G, which is basic for the theory of reductive groups.

9.3.4. Theorem. The identity component V of the intersection of the unipotent radicals of the Borel subgroups containing T coincides with the unipotent radical R_uG.
It is clear that $R_uG \subset V$. To prove the reverse inclusion it suffices to show that V is a normal subgroup of G. Using 8.1.6(iv) we see that it suffices to prove that any Borel subgroup $B \supset T$ normalizes V and using 9.1.2 it follows that we only have to prove that for any codimension 1 subtorus S of T, the group $Z_GS \cap B$ normalizes V. If S is a regular subtorus,

then (by 7.3.5(ii), $Z_G S \subset T.V$ and it is evident that $Z_G S$ normalizes V. Now let S be singular. There is a root $\alpha \in R(G,T)$ such that $Z_G S = G_\alpha$ (notation of 9.3.2) and $\alpha \in R^+(B)$. Moreover, $B_\alpha = G_\alpha \cap B$ is a Borel subgroup of G_α. It is clear that for any Borel subgroup $B' \supset T$ with $\alpha \in R^+(B')$ we have $B_\alpha \subset B'$. Hence B_α lies in the subgroup H_α of 9.3.3, and it follows from 9.3.3 that B_α normalizes V, which is what we wanted.

The theorem has important consequences.

9.3.5. Corollary. *Assume, moreover, that G is reductive.*
(i) *If S is a subtorus of G then $Z_G S$ is connected and reductive;*
(ii) $Z_G T = T$. *In other words, the Cartan subgroups and the maximal tori of G coincide;*
(iii) *The center Z(G) lies in T.*
To prove (i) we may assume $S \subset T$. By the theorem and 7.3.5(ii) we have

$$R_u(Z_G S) = \bigcap_{B \supset T} ((Z_G S) \cap B) \subset \bigcap_{B \supset T} B = R_u G = \{e\},$$

which proves that $Z_G S$ is reductive. That this group is connected was already proved in 9.3.5(i). The particular case $S = T$ gives (ii), and (iii) is then clear since $Z(G) \subset Z_G T$.

Let now G be reductive and put $R = R(G,T)$.

9.3.6. Proposition. (i) *For any $\alpha \in R$ there exists an isomorphism x_α of G_a onto a unique closed subgroup X_α of G such that $t x_\alpha(\xi) t^{-1} = x_\alpha(\alpha(t)\xi)$ $(t \in T, \xi \in k)$. We have (with the notation of 9.1.1)* $\text{Im } dx_\alpha = \mathfrak{g}_\alpha$.

(ii) T *and the* X_α $(\alpha \in R)$ *generate* G.

If $\alpha \in R$ then $G_\alpha = Z_G(\mathrm{Ker}\ \alpha)^0$ is, by 9.3.5(i), a reductive subgroup, of semi-simple rank 1. The existence of x_α and the last statement of (i) then follow from 8.2.7(iii). If X_α is as stated, then $X_\alpha \subset G_\alpha$, and the uniqueness statement of (i) also follows from 8.2.7(iii).

(ii) is a consequence of 9.1.2.

9.3.7. *Corollary. The roots in* R *are the non-zero weights of* T *in* $\mathfrak{g}$. *For any* $\alpha \in R$ *we have* $\dim \mathfrak{g}_\alpha = 1$.

9.3.8. *Corollary. Let* $B \supset T$ *be a Borel subgroup, let* $\alpha \in R$.
(i) *The following properties are* ~~eig~~*envalent*: (a) $\alpha \in R^+(B)$, (b) $X_\alpha \subset B$, (c) $\mathfrak{g}_\alpha \subset \mathrm{Lie}\ B$; equi
(ii) $\dim B = \dim T + \frac{1}{2}|R|$, $\dim G = \dim T + |R|$.

Here $|R|$ denotes the number of elements of the set R.

(i) follows readily from the definition of $R^+(B)$ given in 9.3.1 (check this), and (ii) is a consequence of (i).

9.3.9. *Exercises*. G is a connected reductive linear algebraic group and T a maximal torus of G.

(1) If X is a 1-dimensional connected unipotent subgroup of X which is normalized by T then there is $\alpha \in R$ such that $X = X_\alpha$ (as in 9.3.6).

(2) (a) Let $t \in T$. The connected centralizer $Z_G(t)^0$ is generated by T and the X_α with $\alpha(t) = 1$ (Hint: use 4.4.4(ii)).

(b) The centralizer of a semi-simple element of G is reductive.

(3) Let $G = \mathbb{GL}_n$, $T = \mathbb{D}_n$ (see 2.1.3).

(a) Define characters α_{ij} of T by $\alpha_{ij}(\mathrm{diag}(x_1,\dots,x_n)) = x_i x_j^{-1}$ $(1 \leq i,j \leq n, i \neq j)$. Then α_{ij} is a root of $R(G,T)$.

(b) $\mathfrak{g} = \mathfrak{gl}_n$ is the direct sum of $\mathfrak{t}$ and the subspaces $\mathfrak{g}_{\alpha_{ij}}$ $(i \neq j)$. Show that G is reductive.

(c) The root datum $\Psi(G,T)$ is isomorphic to the following one $(X,R,X^\vee,R^\vee)$, with $X = X^\vee = \mathbb{Z}^n$ (the pairing being the standard one), $R = R^\vee = \{\varepsilon_i - \varepsilon_j \mid i \neq j\}$, where (ε_i) is the canonical basis.

(d) Put $G_1 = \mathbb{SL}_n \subset G$, $T_1 = T \cap G$, then G_1 is a connected semi-simple group and T_1 a maximal torus of G_1. The root datum $\Psi(G_1,T_1)$ is isomorphic to $(X_1,R_1,X_1^\vee,R_1^\vee)$, where $X_1 = X/\mathbb{Z}(\varepsilon_1+\dots+\varepsilon_n)$, $X_1^\vee = \{(x_1,\dots,x_n) \in X^\vee \mid \sum_1^n x_i = 0\}$, notations being as in (c). If π is the canonical map $X \to X_1$, then $R_1 = \pi R$, $R_1^\vee = R^\vee \subset X_1^\vee$.

(e) Let $Z \subset \mathbb{GL}_n$ be the subgroup of scalar multiples of the identity, which is the center of $\mathbb{GL}_n$. Put $G_2 = G/Z$, $T_2 = T/Z$, then G_2 is a connected semi-simple group with maximal torus T_2 and the root datum $\Psi(G_2,T_2)$ is isomorphic to $(X_1^\vee,R_1^\vee,X_1,R_1)$ (notations of (d)).

(4) (char $k \neq 2$). Let $V = k^{2n+1}$, and $(\ ,\ )$ the symmetric bilinear form on V defined by $((\xi_0,\dots,\xi_{2n}),(\eta_0,\dots,\eta_{2n})) = \xi_0\eta_0 + \sum_{i=1}^{n}(\xi_i\eta_{n+i}+\xi_{n+i}\eta_i)$. Denote by G the identity component of the group of $t \in GL(V)$ with $(tv,tw) = (v,w)$ $(v,w \in V)$.

(a) Show that G is isomorphic to $\mathbb{SO}_{2n+1}$ and that the group T of transformations

$(\xi_0,\dots,\xi_{2n}) \mapsto (\xi_0, x_1\xi_1,\dots,x_n\xi_n, x_1^{-1}\xi_{n+1},\dots,x_n^{-1}\xi_{2n})$ $(x_i \in k^*)$

is a maximal torus of G. The maps $(x_1,\dots,x_n) \mapsto x_i^{\varepsilon}$ and $(x_1,\dots,x_n) \mapsto x_i^2 x_j^{\eta}$ $(i \neq j, \varepsilon,\eta = \pm 1)$ define roots of $R(G,T)$.

(b) The Lie algebra $\mathfrak{g}$ is the set of all $t \in \text{End}(V)$ such that $(tv,w) + (v,tw) = 0$ if $v,w \in V$ (Hint: use 4.4.4(ii)). $\mathfrak{g}$ is the direct sum of $\mathfrak{t}$ and the $\mathfrak{g}_\alpha$, the α being as in (a). Show that G is semi-simple.

(c) The root datum $\Psi(G,T)$ is isomorphic to $(X,R,X^\vee,R^\vee)$, where $X = X^\vee = \mathbb{Z}^n$ (standard pairing), $R = \{\pm\varepsilon_i, \pm\varepsilon_i \pm \varepsilon_j \mid i \neq j\}$, $R^\vee = \{\pm 2\varepsilon_i, \pm\varepsilon_i \pm \varepsilon_j \mid i \neq j\}$.

(5) (char $k \neq 2$). Let $V = k^{2n}$, and (,) the symmetric bilinear form on V with $((\xi_1,\ldots,\xi_{2n}),(\eta_1,\ldots,\eta_{2n})) = \sum_{n=1}^{n} (\xi_i\eta_{n+i} + \xi_{n+i}\eta_i)$.

Let G be as in the previous exercise, and T the subgroup of transformations $(\xi_1,\ldots,\xi_{2n}) \mapsto (x_1\xi_1,\ldots,x_n\xi_n,x_1^{-1}\xi_{n+1},\ldots,x_n^{-1}\xi_{2n})$. Again, G is semi-simple (isomorphic to $\mathbf{SO}_{2n}$) and T is a maximal torus.

The root datum $\Psi(G,T)$ is isomorphic to $(X,R,X^\vee,R^\vee)$, with $X = X^\vee = \mathbb{Z}^n$, $R = R^\vee = \{\pm\varepsilon_i \pm \varepsilon_j \mid i \neq j\}$.

(6) Let $V = k^{2n}$, and (,) the alternating bilinear form with $((\xi_1,\ldots,\xi_{2n}),(\eta_1,\ldots,\eta_{2n})) = \sum_{i=1}^{n} (\xi_i\eta_{n+i} - \xi_{n+i}\eta_i)$. Let G be the symplectic group $\mathbf{Sp}_{2n}$ (see (6.6(2)). If T is as before, then T is a maximal torus in G.

Show that G is semi-simple. Its root datum is isomorphic to $(X,R,X^\vee,R^\vee)$ with $X = X^\vee = \mathbb{Z}^n$, $R = \{\pm 2\varepsilon_i, \pm\varepsilon_i \pm \varepsilon_j \mid i \neq j\}$, $R^\vee = \{\pm\varepsilon_i, \pm\varepsilon_i \pm \varepsilon_j \mid i \neq j\}$.

Then root systems R introduced in (3),(4),(5),(6) are denoted A_{n-1}, B_n, D_n, C_n, respectively.

9.4. Semi-simple groups.

We close this chapter with some properties of semi-simple groups. The notations are as before.

9.4.1. *Theorem. Let G be semi-simple.*

(i) *The* X_α ($\alpha \in R$) *generate* G;

(ii) $G = (G,G)$;

(iii) *Let* G_1 *be a connected, closed, normal subgroup of* G. *Then* G_1 *is semi-simple. There is a connected, closed, normal subgroup* G_2 *of* G *such that* $(G_1,G_2) = e$, $G_1 \cap G_2$ *is finite and* $G = G_1G_2$;

(iv) *The number of minimal non-trivial connected, closed, normal subgroups of* G *is finite. If* $G_1,\dots,G_r$ *are these groups, then* $(G_i,G_j) = e$ *and* $G_i \cap \prod_{j\neq i} G_j$ *is finite. Moreover* $G = G_1G_2\dots G_r$ *and the* G_i *have no closed normal subgroups of dimension* > 0.

Let G_1 be as in (iii). Its radical is a connected, solvable, closed normal subgroup of G, hence is trivial. So G_1 is also semi-simple. First assume that G_1 is the subgroup generated by the X_α, which is closed, connected by 6.1 , and normal because T and the X_α's generate G (9.3.6(ii)) and normalize G_1. Since $\bigcap_{\alpha\in R} (\mathrm{Ker}\ \alpha)^0$ is a central torus in G this group is trivial. This means that the roots $\alpha \in R$ span a sublattice of X of finite index and also that T is generated by the groups $\alpha^\vee(\mathbb{G}_m)$ (check this).

Since $\alpha^\vee(\mathbb{G}_m)$ lies in the subgroup generated by X_α and $X_{-\alpha}$, we have that $T \subset G_1$, whence (i) (using 9.3.6(ii)).

Let $\alpha \in R, t \in T$. Then

(*) $tx_\alpha(\xi)t^{-1}x_\alpha(-\xi) = x_\alpha((\alpha(t)-1)\xi)$,

From (*) one sees that $X_\alpha \subset (G,G)$, whence (ii).
Next let G_1 be as in (iii). Let T_1 be a maximal torus of G_1, which we may assume to be contained in T. Assume $X_\alpha \not\subset G_1$ and take $t \in T_1$ in (*). It follows that we must have $\alpha(T_1) = 1$. Conversely, if this is so we must have $X_\alpha \not\subset G_1$, by 9.3.5(ii). Let $R_1 = \{\alpha \in R | X_\alpha \subset G_1\}$, $R_2 = R-R_1$. If $R_1 = \phi$ then $T_1 \subset \bigcap_{\alpha \in R} (\operatorname{Ker} \alpha)^0 = e$, whence $G_1 = e$. If $R_2 = \phi$ then $G_1 = G$ by (i). In both cases (iii) holds. Now assume $R_1 \neq \phi$, $R_2 \neq \phi$, and let $\alpha \in R_1, \beta \in R_2$. Then, putting

$$x(\xi) = x_\beta(\eta)x_\alpha(\xi)x_\beta(-\eta),$$

we have $x(\xi) \in G_1$ and $tx(\xi)t^{-1} = x(\alpha(t)\xi)$ $(t \in T_1)$. It follows from 9.3.9(1) that $x(\xi) = x_\alpha(f(\eta)\xi)$, where f is a homomorphism of algebraic groups $\mathbb{G}_a \to \mathbb{G}_m$, which must be trivial. We conclude that $(X_\alpha, X_\beta) = e$. If G_2 is the subgroup generated by the X_β $(\beta \in R_2)$ we have $(G_1,G_2) = e$, $G = G_1G_2$. That $G_1 \cap G_2$ is finite follows from the fact that it is a closed normal subgroup not containing any X_α. This proves (iii), and (iv) is an easy consequence of (iii), the proof of which is left to the reader.

9.4.2. Assuming G semi-simple, let $(X,R,X^\vee,R^\vee)$ be the root datum. As we saw in the proof of 9.4.1, the sublattice Q of X spanned by R has finite index in X. Put $V = Q \otimes_{\mathbb{Z}} \mathbb{R}$, as in 9.1.6, and let $P = \{v \in V | \langle v,R^\vee \rangle \in \mathbb{Z}\}$. Then P is a lattice in V containing Q and it is clear that $Q \subset X \subset P$. So if the root system $R \subset V$ is given there are only finitely many possibilities for X.

G is said to be *adjoint* if $X = Q$ and *simply connected* if $X = P$.

Q is the *root lattice* of R and P the *weight lattice*. The finite abelian group P/Q is the *fundamental group* of R.

9.4.3. *Exercises*. (1) (a) Define the notion of direct sum of two root systems.

(b) A root system is *reducible* if it is a nontrivial direct sum and *irreducible* otherwise. Show that the root system R of G is reducible if and only if G has a proper connected, closed, normal subgroup. Show that the root systems of the groups G_i of 9.4.1(iii) are irreducible.

(2) Let $(X,R,X^\vee,R^\vee)$ be the root datum of a reductive G. Let $X_0 = \{x \in X \mid \langle x,R^\vee \rangle = 0\}$. As before, Q is the ~~weight~~ root lattice.

(a) (G,G) is adjoint if and only if $X = Q \oplus X_0$.

(b) (G,G) is simply connected if and only if $P \subset X+(X_0 \otimes Q)$.

(3) Determine root lattice, weight lattice and fundamental group for each of the root systems A_{n-1}, B_n, C_n, D_n, introduced in 9.3.9.

Notes.

Most of the material in this chapter is more or less standard. The definition of roots, given in 9.1.4, is Chevalley's original one [11 , exp.12]. We have proved, along the way, the properties of root systems which are needed in this chapter.

The notion of a root datum (taken from [14, exp. XXI]), has been introduced as soon as possible. It is needed for a clean formulation of the existence and uniqueness theorems for reductive groups (12.1 and 11.4.2), and also for the theory of automorphic L-functions (see [4]).

10. Further properties of reductive groups.

In this chapter G is a reductive connected linear algebraic group, T a maximal torus of G and $B \supset T$ a Borel subgroup. We denote by $R = R(G,T)$ the root system with respect to T and we put $R^+ = R^+(B)$. The notion of positive root is relative to R^+ (see 9.2.2). The Weyl group is $W = N_G T/T$ (see 9.3.5(ii)). If $\alpha \in R$, let x_α and X_α be as in 9.3.6.

10.1. Borel groups and systems of positive roots.

10.1.1. Proposition. Let $(\alpha_1, \ldots, \alpha_n)$ be the set of elements of R^+, in some order. The morphism $\phi: \mathbb{G}_a^n \to B_u$ with $\phi(\xi_1, \ldots, \xi_n) = x_{\alpha_1}(\xi_1) \ldots x_{\alpha_n}(\xi_n)$ is an isomorphism of varieties. In particular, B_u is generated by the X_α with $\alpha > 0$.

It is convenient to prove, instead of 10.1.1, the following more general result.

10.1.2. Lemma. Let H be a connected solvable linear algebraic group, let S be a maximal torus of H and $U = H_u$. Assume that there is a set of isomorphisms x_i $(1 \leqslant i \leqslant m)$ of $\mathbb{G}_a$ onto closed subgroups of H such that

(a) there exist distinct characters β_i $(1 \leqslant i \leqslant m)$ of S with $s x_i(\xi) s^{-1} = x_i(\beta_i(s)\xi)$ $(s \in S, \xi \in k)$;

(b) $\mathfrak{u}$ is the direct sum of the weight spaces $\mathfrak{u}_{\beta_i}$ $(1 \leqslant i \leqslant m)$ and these are all 1-dimensional.

Then the morphism $\psi: (\mathbb{G}_a)^m \to U$ with $\psi(\xi_1, \ldots, \xi_m) = x_1(\xi_1) \ldots x_m(\xi_m)$ is an isomorphism of varieties.

The proof of 10.1.2 is by induction on m. If m = 1 there is

no problem (since then $U = \operatorname{Im} x_1$). If $m > 1$, let $N \subset U$ be a normal subgroup of H contained in the center of U and isomorphic to $\mathbb{G}_a$ (see 6.10). Then Lie N must be one of the weight spaces, say $\mathfrak{u}_{\beta_j}$. It follows that $N = \operatorname{Im} x_j$ (look at $Z_H(\operatorname{Ker} \beta_j)^0$). By induction we may assume the result to be true for H/N. Since $\operatorname{Im} x_j$ is central in U, it follows immediately that the morphism ψ is bijective. That it is an isomorphism then comes from the fact that the tangent map $(d\psi)_{(0,\ldots,0)}$ is surjective, which is a consequence of (b). It is left to the reader to check that 10.1.2 implies 10.1.1.

We next discuss some properties of the "additive 1-parameter subgroups" x_α. First notice that x_α is not uniquely determined by the properties of 9.3.6(i). In fact, α being given, x'_α has the same properties if and only if there is $c \in k^*$ such that $x'_\alpha(\xi) = x_\alpha(c\xi)$, as follows from 8.2.7(iii). The following result is now obvious.

10.1.3. <u>Lemma</u>. <u>Let</u> $n \in N_G T$ <u>and put</u> $w = nT$. <u>There exist</u> $c_{n,\alpha}$ $(\alpha \in R)$ <u>in</u> k^* <u>such that</u> $n x_\alpha(\xi) n^{-1} = x_{w.\alpha}(c_{n,\alpha}\xi)$ $(\xi \in k)$.

The next proposition gives information about commutators $(x_\alpha(\xi), x_\beta(\eta))$.

10.1.4. <u>Proposition</u>. <u>Let</u> $\alpha, \beta \in R$, $\alpha \neq \pm \beta$. <u>There exist constants</u> $c_{\alpha,\beta;i,j} \in k$ <u>such that</u>

$$(x_\alpha(\xi), x_\beta(\eta)) = \prod_{\substack{i\alpha + j\beta \in R \\ i,j > 0}} x_{i\alpha+j\beta}(c_{\alpha,\beta;i,j}\xi^i\eta^j),$$

the product in the right-hand side being taken in some preassigned order.

From 9.2.3 and 10.1.3 we see that it is sufficient to prove this if $\alpha > 0$, $\beta > 0$. In that case we have $X_\alpha \subset B, X_\beta \subset B$ (10.1.1). By 10.1.1 we can write

$$(x_\alpha(\xi), x_\beta(\eta)) = \prod_{\gamma > 0} x_\gamma(P_\gamma(\xi,\eta)),$$

where $P_\gamma \in k[T,U]$, the product being taken in some preassigned order. Applying the inner automorphism defined by $t \in T$ we see that

$$(x_\alpha(\alpha(t)\xi), x_\beta(\beta(t)\xi)) = \prod_{\gamma > 0} x_\gamma(\gamma(t)P_\gamma(\xi,\eta)),$$

from which we conclude that

$$P_\gamma(\alpha(t)\xi, \beta(t)\eta) = \gamma(t)P_\gamma(\xi,\eta).$$

Write $P_\gamma(T,U) = \sum_{i,j \geqslant 0} c_{ij}T^iU^j$, then

$$\gamma(t) \sum_{i,j \geqslant 0} c_{ij}\xi^i\eta^j = \sum_{i,j \geqslant 0} c_{ij}\alpha(t)^i\beta(t)^j\xi^i\eta^j,$$

and the linear independence of characters implies that only those γ can occur which have the form (written additively) $i\alpha+j\beta$, where $i \geqslant 0$, $j \geqslant 0$. Since α and β are independent (9.1.7) it follows that i and j are unique, if they exist. To conclude the proof we have to show that we must have $i > 0$, $j > 0$. Suppose, for example, that we had a formula of the required type, in which in the right-hand side a term occurred with $j = 0$. Then we had $i = 1$, and the formula had the form

$$(x_\alpha(\xi),x_\beta(\xi)) = x_\alpha(c\xi) \prod_{i>0} x_{i\alpha+j\beta}(c_{\alpha,\beta;i,j}\xi^i\eta^j),$$

with $c \neq 0$. But then it would follow that we had a similar formula for all repeated commutators:

$$(..(x_\alpha(\xi),x_\beta(\eta))..)x_\beta(\eta)) =$$

$$x_\alpha(d\xi) \prod_{i>0} x_{i\alpha+j\beta}(d_{\alpha,\beta;i,j}\xi^i\eta^j),$$

with $d \neq 0$, which contradicts the nilpotency of B_u.

10.1.5. Theorem. Let $\tilde{R}^+$ be an arbitrary system of positive roots in R.

(i) T and the X_α with $\alpha \in \tilde{R}^+$ generate a Borel subgroup of G;

(ii) There is a unique $w \in W$ such that $\tilde{R}^+ = w.R^+$.

Denote by $\tilde{R}^+_n$ the set of roots in $\tilde{R}^+$ which are linear combinations, with strictly positive coefficients, of at least n other elements of $\tilde{R}^+$ ($n \geq 1$) and let U_n be the subgroup of G generated by the x_α with $\alpha \in \tilde{R}^+_n$. By descending induction on n one shows, that U_n is a closed, connected, unipotent subgroup of G, normalized by T, whose dimension equals $|\tilde{R}^+_n|$. Hence $\tilde{B} = TU_1$ is a closed, connected, solvable subgroup of G, whose dimension is $\dim T + \frac{1}{2}|R|$. We conclude from 9.3.8(ii) that $\tilde{B}$ must be a Borel subgroup of G, containing T. This proves (i). (ii) is a consequence of (i) and 8.1.5(ii).

As in 9.1.8, (,) is a positive definite symmetric bilinear form on $X_{\mathbb{R}}$ which is W-invariant.

10.1.6. Corollary. There is $x \in X_{\mathbb{R}}$ such that $\tilde{R}^+ = \{\alpha \in R| (x,\alpha) > 0\}$. We can take for x any vector in $X_{\mathbb{R}}$ such that

$(x,\alpha) > 0$ <u>for all</u> $\alpha \in \tilde{R}^+$.

This is true for R^+ (see the proof of 9.3.2). The first assertion then follows from 10.1.5(ii). The last one is obvious from the fact that $R = \tilde{R}^+ \cup (-\tilde{R}^+)$.

Denote by $\mathcal{B}_T$ the set of Borel subgroups of G containing T. We say that $B_1, B_2 \in \mathcal{B}_T$ are <u>adjacent</u> if $\dim B_1 \cap B_2 = \dim B_1 - 1 = \dim B_2 - 1$.

10.1.7. <u>Proposition</u>. <u>If</u> $B, B' \in \mathcal{B}_T$ <u>there is a chain</u> $B = B_0, B_1, \ldots, B_h = B'$ <u>of elements of</u> $\mathcal{B}_T$ <u>such that</u> B_i <u>and</u> B_{i+1} <u>are adjacent</u> $(0 \leqslant i \leqslant h-1)$.

By 10.1.6 there exist $x, y \in X_{\mathbb{R}}$ such that $R^+ = R^+(B)$ $(R^+(B'))$ is the set of $\alpha \in R$ with $(x,\alpha) > 0$ (resp. $(y,\alpha) > 0$). Clearly, the x with this property form an open subset of $X_{\mathbb{R}}$ (with its Euclidean topology). By varying x a little we may assume that

$$(\alpha,x)^{-1}(\beta,x) \neq (\alpha,y)^{-1}(\beta,y),$$

for all $\alpha, \beta \in R$ with $\alpha \neq \pm\beta$.

For $0 \leqslant t \leqslant 1$ put $x(t) = (1-t)x+ty$. If $(x(t),\alpha) = 0$ for some t then $(x,\alpha) \neq (y,\alpha)$ and $t = t_\alpha = ((x,\alpha)-(y,\alpha))^{-1}(x,\alpha)$. From our choice of x we see that $t_\alpha \neq t_\beta$ if $\alpha \neq \pm\beta$. It follows that there exist numbers $0 = t_0 < t_1 < \ldots < t_h = 1$ such that each t_i is a t_α $(0 < i < h)$ and that all $(x(t),\alpha)$ $(\alpha \in R^+)$ have a constant sign if t lies in the interval $I_i = (t_i, t_{i+1})$ $(0 \leqslant i \leqslant h-1)$. Moreover the sign distributions for I_i and I_{i+1} differ for only one $\alpha \in R^+$.

Let $R_i^+ = \{\alpha \in R | (x(t),\alpha) > 0, t \in I_i\}$ and let $B_i \in \mathcal{B}_T$ be Borel subgroup with $R^+(B_i) = R_i^+$ (10.1.5). Then B_i and B_{i+1} are adjacent.

Recall that W operates simply transitively on $\mathcal{B}_T$ (8.1.5(ii)).

10.1.8. Lemma. *Let B' be adjacent to B. There is a unique* $\alpha \in R^+$ *such that* $B' = s_\alpha . B$.
We have $(B \cap B')^0 = T(B_u \cap B_u')^0$ and $(B_u \cap B_u')^0$ has codimension 1 in B_u and B_u', hence is normal in these groups (6.6(3)). It follows that $(B_u \cap B_u')^0$ is a unipotent closed normal subgroup of the parabolic subgroup P generated by B and B'. Then P is of semi-simple rank 1 (check this) and the statement of the lemma follows from the discussion in 8.2.

Let $D = D(B)$ be the set of roots $\alpha \in R^+ = R^+(B)$ such that $s_\alpha . B$ and B are adjacent. We call D the *basis* of R^+; this name is explained by 10.1.10(iii). The $\alpha \in D$ are called the *simple roots* of R^+. Let $S = S(B)$ be the set of reflections s_α ($\alpha \in D$).
It is clear that $D(w.B) = w.D(B)$, $S(w.B) = wSw^{-1}$.

10.1.9. Lemma. (i) *If* $\alpha \in D$ *then* s_α *permutes the elements of* $R^+ - \{\alpha\}$;
(ii) *If* $\alpha, \beta \in D$, $\alpha \neq \beta$ then $(\alpha,\beta) \leq 0$.
Since $s_\alpha \alpha = -\alpha$, (i) is obvious from the definition of D.
Assume $\alpha, \beta \in D$, $\alpha \neq \beta$. Then, x being as in the proof of 10.1.7 we have $(x, \langle\alpha,\beta^\vee\rangle s_\alpha\beta + s_\beta\alpha) = (1 - \langle\alpha,\beta^\vee\rangle\langle\beta,\alpha^\vee\rangle)(x,\alpha)$.
If $(\alpha,\beta) > 0$, the left-hand side is > 0 since $s_\alpha\beta$ and $s_\beta\alpha$ lie

in R^+ by (i) and the right-hand side is < 0 (see 9.2.1). This proves (ii).

The following theorem gives the main properties of the basis D and the corresponding set of reflections S.

10.1.10. Theorem. (i) *S generates* W;

(ii) R = W.D;

(iii) *The simple roots are linearly independent. Each element of* R^+ *is a linear combination* $\sum_{\alpha \in D} n_\alpha \alpha$, *with* $n_\alpha \in \mathbb{Z}$, $n_\alpha \geq 0$. *These two properties characterize the set of simple roots.*

Let $w \in W$. Applying 10.1.7 we find a chain $B = B_0, B_1, \ldots, B_h = w.B$, with B_i and B_{i+1} adjacent. Write $B_{h-1} = w'.B$. By induction on h we may assume that w' is a product of elements of S. Since w'.B and w.B are adjacent, the same is true for $(w')^{-1}w.B$ and B, whence $(w')^{-1}w \in S$. This implies (i).

Now we prove linear independence of the simple roots. Assume $\sum_{\alpha \in D} n_\alpha \alpha = 0$ is a dependence relation. We may assume that $n_\alpha > 0$ for some α. Let $D_1 = \{\alpha \in D | n_\alpha > 0\}$ and rewrite the relation as

$$\sum_{\alpha \in D_1} n_\alpha \alpha = \sum_{\beta \in D - D_1} m_\beta \beta,$$

with $n_\alpha > 0$, $m_\beta \geq 0$.

If $t = \sum_{\alpha \in D_1} n_\alpha \alpha$, then $(t,t) = \sum_{\substack{\alpha \in D_1 \\ \beta \in D - D_1}} n_\alpha m_\beta (\alpha, \beta) \leq 0$ (by 10.1.9(ii)) whence $t = 0$. But, x being as before, $(x,t) > 0$, which is a contradiction. This proves independence.

Let $\alpha \in R^+ - D$ and $\beta \in D$. Then $s_\beta\alpha = \alpha - \langle\alpha,\beta^\vee\rangle\beta$ and $(x,s_\beta\alpha) = (x,\alpha) - \langle\alpha,\beta^\vee\rangle(x,\beta)$. By 10.1.9(i) we have $s_\beta\alpha \in R^+$. If $\langle\alpha,\beta^\vee\rangle > 0$ we have $0 < (x,s_\beta\alpha) < (x,\beta)$. If we know already that $s_\beta\alpha \in w.D$ and is a positive integral linear combination of elements of D, then the same will be true for α. It follows (check this) that (ii) and the first assertion of (iii) are true if we establish the following: if $\alpha \in R^+ - D$ there is $\beta \in D$ such that $\langle\alpha,\beta^\vee\rangle > 0$.

Let $\alpha \in R$ and assume that $\langle\alpha,\beta^\vee\rangle \leq 0$ for all $\beta \in D$. We claim that this implies $\alpha \in -R^+$, which will establish the previous assertion. Write

$$\alpha = \sum_{\beta \in D} \xi_\beta \beta + t,$$

with $\xi_\beta \in \mathbb{R}$, and $t \in X_R$ such that $(t,D) = 0$.

From (i) it follows that W stabilizes the subspace of $X_{\mathbb{R}}$ spanned by D and that $w.t = t$ for all $w \in W$. Applying this with $w = s_\alpha$ we see that $t = 0$. Now put $u = \alpha - \sum_{\xi_\beta < 0} \xi_\beta \beta$. Then

$$(u,u) = (\alpha - \sum_{\xi_\beta < 0} \xi_\beta\beta, \sum_{\xi_\beta \geq 0} \xi_\beta\beta) \leq 0,$$

whence $u = 0$.

This shows that α is a negative linear combination of the elements of D, whence $\alpha \in -R^+$, proving our claim. To prove the uniqueness statement observe that $\alpha \in R^+$ is simple if and only if it cannot be written as the sum of two positive roots, which follows from the preceding argument.

10.1.11. <u>Corollary</u>. G <u>is generated</u> by T <u>and the groups</u> $X_{\pm\alpha}$ ($\alpha \in D$).

Let G_1 be the subgroup so generated, it is closed connected subgroup (6.1). Its Lie algebra contains $\mathfrak{t}$ and the subspaces $\mathfrak{g}_{\pm\alpha}$ ($\alpha \in D$). It is readily checked that 10.1.10(i), (ii) imply that Lie G_1 contains <u>all</u> $\mathfrak{g}_\alpha$ ($\alpha \in R$), whence $G_1 = G$.

10.1.12. <u>Exercises</u>. The notations are as before. R is a root system and D a basis.

(1) R is one of the root systems A_{n-1}, B_n, C_n, D_n. The notations are as in 9.3.9.

(a) The set D described below is a basis of R.

A_{n-1}: $S = \{\varepsilon_1-\varepsilon_2, \varepsilon_2-\varepsilon_3, \dots, \varepsilon_{n-1}-\varepsilon_n\}$,

B_n : $S = \{\varepsilon_1-\varepsilon_2, \varepsilon_2-\varepsilon_3, \dots, \varepsilon_{n-1}-\varepsilon_n, \varepsilon_n\}$,

C_n : $S = \{\varepsilon_1-\varepsilon_2, \varepsilon_2-\varepsilon_3, \dots, \varepsilon_{n-1}-\varepsilon_n, 2\varepsilon_n\}$,

D_n : $S = \{\varepsilon_1-\varepsilon_2, \varepsilon_2-\varepsilon_3, \dots, \varepsilon_{n-1}-\varepsilon_n, \varepsilon_{n-1}+\varepsilon_n\}$.

(b) R is irreducible.

(c) The Weyl group is as follows.

A_{n-1}: $W = S_n$, acting as the group of permutations of the basis $(\varepsilon_i)_{1\leqslant i\leqslant n}$; B_n, C_n : W is the group of linear transformations of $V = \mathbb{R}^n$ with $\varepsilon_i \mapsto \eta_i\varepsilon_{\sigma i}$, where $\sigma \in S_n$ and $\eta_i = \pm 1$; D_n: W is the subgroup of the preceding group whose elements satisfy $\prod_{i=1}^{n} \eta_i = 1$.

(2) R is reducible (9.4.3(1)) if and only there is a non-trivial decomposition $D = D_1 \cup D_2$ such that $(\alpha,\beta) = 0$ if $\alpha \in D_1$, $\beta \in D_2$.

(3) Let P be the weight lattice (9.4.2). Show that P has a

basis $(\omega_{\alpha})_{\alpha \in D}$, where $\langle \omega_\alpha, \beta^\vee \rangle = \delta_{\alpha\beta}$ (Krnecker symbol). We have $s_\alpha \omega_\beta = \omega_\beta - \delta_{\alpha\beta}\alpha$ (the ω_α are the _fundamental_ _weights_).

10.2. _Bruhat's lemma_.

10.2.1. We keep the notations introduced in the preceding section, in particular G is supposed to be reductive. If $w \in W$, let $R(w) = \{\alpha \in R^+ | w\alpha \in -R^+\}$. By 10.1.9(i) we then have $R(s_\alpha) = \{\alpha\}$, if $\alpha \in D$. This implies immediately that, for $\alpha \in D$,

$$(1) \quad \begin{cases} R(ws_\alpha) = s_\alpha R(w) \cup \{\alpha\} \text{ if } w\alpha > 0, \\ R(ws_\alpha) = s_\alpha(R(w) - \{\alpha\}) \text{ if } w\alpha < 0. \end{cases}$$

If $w \in W$ we can write $w = s_1 \ldots s_h$ with $s_i \in S$ (10.1.10(i)). If h is as small as possible we call this a _shortest_ _expression_ of w. Then h is the _length_ $l(w)$ of w (with respect to S,D or R^+). Notice that $l(w^{-1}) = l(w)$.

10.2.2. _Lemma_. _Let_ $w = s_1 \ldots s_h$ _be a shortest expression, and_ $s_i = s_{\alpha_i}$ $(\alpha_i \in D)$.

(i) $R(w) = \{\alpha_h, s_h\alpha_{h-1}, \ldots, s_h \ldots s_2\alpha_1\}$;

(ii) _The number of elements of_ $R(w)$ _equals the length_ $l(w)$;

(iii) _If_ $\alpha \in D$, _then_ $l(ws_\alpha) = l(w)+1$ _if_ $w\alpha > 0$ _and_ $l(ws_\alpha) = l(w)-1$ _if_ $w\alpha < 0$ _and similarly for_ $l(s_\alpha w)$;

(iv) _If_ $\alpha \in D$ _and_ $w\alpha < 0$, _there is a shortest expression_ $w = s'_1 \ldots s'_h$ _with_ $s'_h = s_\alpha$;

(v) _If_ $w = s'_1 \ldots s'_h$ _is any shortest expression, there is an_ i $(1 \leqslant i \leqslant h)$ _with_ $s_1 \ldots s_{i-1}s_{i+1} \ldots s_h = s'_1 \ldots s'_{h-1}$.

(i) is true if $h = 1$. Now let $h > 1$ and write $w' = s_1 \dots s_{h-1}$. By induction we may assume that $R(w') = \{\alpha_{h-1}, s_{h-1}\alpha_{h-2}, \dots, s_{h-1} \dots s_2\alpha_1\}$ and (i) will follow from (1) if we show that $w'\alpha_h > 0$. Now if $w'\alpha_h < 0$, we would have $\alpha_h = s_{h-1} \dots s_{i+1}\alpha_i$ for some i, whence $s_h = s_{\alpha_h} = s_{h-1} \dots s_{i+1}s_i s_{i+1} \dots s_{h-1}$ and $w = s_1 \dots s_h = s_1 \dots s_{i-1}s_{i+1} \dots s_{h-1}$, which contradicts the minimality of h. This establishes (i). Clearly, (ii) is a consequence of (i). The first part of (iii) follows from (i) and (1). For the last statement of (iii) use that $l(w) = l(w^{-1})$.
Let $\alpha \in D$ and $w\alpha < 0$. By (i) we have $\alpha = s_h \dots s_{i+1}\alpha_i$ for some i, and $s_\alpha = s_h \dots s_{i+1}s_i s_{i+1} \dots s_h$. So $ws_\alpha = s_1 \dots s_{i-1}s_{i+1} \dots s_h$, whence (iv). Finally, in the situation of (v), if $s'_h = s_\alpha$, then $w\alpha < 0$ by (i) and (v) follows from the formula just established.

As before, S is the set of reflections s_α ($\alpha \in D$). If $s,t \in S$ and $s \neq t$ denote by $m(s,t)$ the order of st in W. We put $m(\alpha,\beta) = m(s_\alpha, s_\beta)$ ($\alpha,\beta \in D$, $\alpha \neq \beta$). Notice that $m(s,t) = m(t,s)$.

10.2.3. Proposition. *Let μ be a map of S into a multiplicative monoid with 1 with the following property: if $s,t \in S$, $s \neq t$ then*

$$\mu(s)\mu(t)\mu(s)\dots = \mu(t)\mu(s)\mu(t)\dots,$$

where in both sides the product has $m(s,t)$ factors. Then there exists a unique extension of μ to W such that _for any shortest expression_ $w = s_1 \dots s_h$ _we have_

(2) $\mu(w) = \mu(s_1)\dots\mu(s_h)$.

We have to show that the right-hand side of (2) is independent of the choice of the shortest expression. We use induction on $h = l(w)$, starting with $h = 1$. Assume that $w = s_1 \dots s_h = s'_1 \dots s'_h$ are two shortest expressions. By 10.2.2(v) there is i such that $s_1 \dots s_{i-1} s_{i+1} \dots s_h = s'_1 \dots s'_{h-1}$. If $i > 1$ we have, using induction, $\mu(s'_1)\dots\mu(s'_h) = \mu(s_1)\dots\mu(s_{i-1})\mu(s_{i+1})\dots\mu(s_h)\mu(s'_h) = \mu_1(s_1)\dots\mu(s_{i-1})\mu(s_i)\dots\mu(s_h)$, since $s_{i+1}\dots s_h s'_h = s_i \dots s_h$ has length $< h$. If $i = 1$ then $w = s_2 \dots s_h s'_h$ and $\mu(s'_1)\dots\mu(s'_h) = \mu(s_2)\dots\mu(s_h)\mu(s'_h)$. By symmetry, we may also assume that $w = s'_2 \dots s'_h s_h$ and $\mu(s_1)\dots\mu(s_h) = \mu(s'_2)\dots\mu(s'_h)\mu(s_h)$. This reduces the proof to the case that the shortest expressions are of the form $w = w_1 s's = w'_1 ss'$, where $l(w_1) = l(w'_1) = h-2$ and $s, s' \in S$, $s \neq s'$. Proceeding as before with the new reduced expressions we see that we can reduce the proof to the case that the shortest expressions are $w = w_i s's \dots s's = w'_i ss' \dots ss'$, with $l(w_i) = l(w'_i) = h-2i$. We end up with the case that the shortes expressions are $w = s'ss' = ss's\dots$, where $s, s' \in S$, $s \neq s'$. But this is the case covered by the assumption (check this). The proposition follows.

10.2.4. _Proposition_. _The set of generators_ S _and the relations_ $s^2 = 1$, $(st)^{m(s,t)} = 1$ $(s,t \in S, s \neq t)$ _form a_ presentation of the Weyl group W.

Let $\tilde{W}$ be the group with generators $\tilde{s}$ ($s \in S$) and relations $\tilde{s}^2 = 1$, $(\tilde{s}\tilde{t})^{m(s,t)} = 1$ ($s,t \in S, s \neq t$) and $\pi: \tilde{W} \to W$ the obvious surjective homomorphism. By the preceding result there is a map $\mu: W \to \tilde{W}$ with $\mu(w) = \mu(s_1)\dots\mu(s_h)$, if $w = s_1\dots s_h$ is a shortest expression. It is clear that $\pi\circ\phi = \mathrm{id}$ and $\phi\circ\pi = \mathrm{id}$. Consequently π is bijective, whence the result.

Let $w_0 \in W$ be the element with $w_0R^+ = -R^+$. It exists and is unique by 10.1.5(ii). It follows from 10.2.2(ii) that w_0 is the element of W with maximal length. We put $B_u = U$.

10.2.5. Lemma. (i) *For any $w \in W$, the groups X_α with $\alpha \in R(w)$ generate a closed connected subgroup U_w of U, which is normalized by T;*

(ii) *The product morphism $U_w \times U_{w_0w} \to U$ is an isomorphism of varieties.*

(i) follows from 10.1.4, taking into account 6.2, and (ii) is then a consequence of 10.1.1 (check this).

Denote by $\dot{w}$ a representative in N_GT of $w \in W$. We write $C(w)$ for the double coset $B\dot{w}B = \{b\dot{w}b' \mid b,b' \in B\}$. It is readily seen that $C(w)$ is independent of the choice of $\dot{w}$. Since $C(w)$ can be viewed as an orbit of $B \times B$, acting on G, it is (by 4.3.1(i)) a locally closed subvariety of G (see 1.9.6(2)).

10.2.6. Lemma. *Let $w = s_1\dots s_h$ be a shortest expression, with $s_i = s_{\alpha_i}$ ($\alpha_i \in D$). The morphism $\phi: \mathbb{A}^h \times B \to G$ with*

$$\phi(\xi_1,\dots,\xi_h;b) = x_{\alpha_1}(\xi_1)\dot{s}_1x_{\alpha_2}(\xi_2)\dot{s}_2\dots x_{\alpha_h}(\xi_h)\dot{s}_hb$$

defines an isomorphism of varieties $A^h \times B \tilde{\to} C(w)$.

We have $C(w) = B\dot{w}B = U\dot{w}B = U_{w^{-1}}U_{w_0w^{-1}}\dot{w}B$ (by 10.2.5).

Since $\dot{w}^{-1}U_{w_0w^{-1}}\dot{w} \subset B$ it follows that $C(w) = U_{w^{-1}}\dot{w}B$. Since $R(w^{-1}) = \{\alpha_1, s_1\alpha_2, \ldots, s_1\ldots s_{h-1}\alpha_h\}$ by 10.2.2(i), we conclude from 10.1.2 and 10.2.5 that

$$U_{w^{-1}} = X_{\alpha_1}X_{s_1\alpha_2}\ldots X_{s_1\ldots s_{h-1}\alpha_h} =$$

$$X_{\alpha_1}\dot{s}_1(X_{\alpha_2}\ldots X_{s_2\ldots s_{h-1}\alpha_h})\dot{s}_1^{-1},$$

whence

$$U_{w^{-1}} = X_{\alpha_1}\dot{s}_1 U_{w^{-1}s_1}\dot{s}_1^{-1}.$$

By an induction on h we may assume that the assertion is true for $s_1w = s_2\ldots s_h$. It then follows that

$$C(w) = U_{w^{-1}}\dot{w}B = X_{\alpha_1}\dot{s}_1 U_{w^{-1}s_1}\dot{s}_1^{-1}wB = X_{\alpha_1}\dot{s}_1 C(s_1w).$$

This implies that ϕ is surjective. It is readily seen that it is bijective, using that $U_{w^{-1}} \cap \dot{w}B = \{e\}$.

To prove, finally, that ϕ is an isomorphism it suffices to show that the morphism $U_{w^{-1}} \times B \to \dot{w}^{-1}C(w)$ sending (u,b) to $\dot{w}^{-1}u\dot{w}b$ is an isomorphism. This is easily done by determining The tangent map at (e,e). We leave the details to the reader.

10.2.7. *Theorem.* ("*Bruhat's lemma*"). G *is the disjoint union of the double cosets* C(w) $(w \in W)$.

Let $G_1 = \bigcup_{w\in W} C(w)$. It is clear that $BG_1 = G_1$. We claim that

$\dot{s}_\alpha G_1 = G_1$ ($\alpha \in D$). If this has been established it follows from 10.1.11 that G_1 is stable under left multiplications, whence $G_1 = G$.

Consider $\dot{s}_\alpha C(w)$. If $l(s_\alpha w) = l(w)+1$ it follows from 10.2.6 that $\dot{s}_\alpha C(w) \subset C(s_\alpha w)$. If $l(s_\alpha w) = l(w)-1$ we can write (by 10.2.2(iii), applied to w^{-1}) $w = s_\alpha w'$ with $l(w') = l(w)-1$. By 10.2.6 we have

$$C(w) = X_\alpha \dot{s}_\alpha C(w'),$$

whence

$$\dot{s}_\alpha C(w) = \dot{s}_\alpha X_\alpha \dot{s}_\alpha C(w')$$

Let $G_\alpha = Z_G(\mathrm{Ker}\ \alpha)^0$, as in 9.1.4, then $s_\alpha X_\alpha s_\alpha \subset G_\alpha$. It follows from 8.2.2(i) (using 7.3.5(ii)) that $G_\alpha \subset (G_\alpha \cap B) \cup X_\alpha \dot{s}_\alpha (G_\alpha \cap B)$, whence

$$\dot{s}_\alpha C(w) \subset G_\alpha C(w') \subset C(w') \cup C(w),$$

which establishes the claim made above.

To complete the proof we have to show that $C(w) \cap C(w') \neq \phi$ implies $w = w'$. We prove by induction on $l(w)$ the equivalent statement: if $w, w' \in W$, $l(w) \leqslant l(w')$ and $C(w) = C(w')$ then $w = w'$.

If $l(w) = 0$ then $C(w) = B \subset C(w')$ and the assertion follows because $N_G T \cap B = T$ (see 6.12(ii)). Now assume $h = l(w) > 0$ and write $w = s_\alpha w$, with $l(w_1) = h-1$. We have seen that then $C(w) = \dot{s}_\alpha C(w_1)$. So $C(w_1) = \dot{s}_\alpha^{-1} C(w') = \dot{s}_\alpha C(w')$. If $l(s_\alpha w') = l(w') + 1$ it follows that $C(w_1) = C(s_\alpha w')$, whence $w_1 = s_\alpha w'$

or $w = w'$, by induction. If $l(s_\alpha w') = l(w')-1$ we have (see the first part of the proof) $\dot{s}_\alpha C(w') \subset C(w') \cup C(s_\alpha w')$. By induction it follows that either $w = w'$ or $w = s_\alpha w'$. The second alternative is impossible because $l(w) \leq l(w')$. This concludes the proof of 10.2.7.

10.2.8. *Corollary*. *We have*

$$\begin{cases} C(s_\alpha)C(w) = C(s_\alpha w) & , (l(s_\alpha w) = l(w)+1), \\ C(s_\alpha)C(w) = C(s_\alpha w) \cup C(w), & (l(s_\alpha w) = l(w)-1). \end{cases}$$

This readily follows from what was established in the first part of the proof.

10.2.9. *Corollary*. *Let* B' *be a Borel subgroup of* G. *Then* $B \cap B'$ *contains a maximal torus*.

Let $B' = gBg^{-1}$ (7.2.6(iii)), and write $g = b\dot{w}b'$, with $w \in W$, $b,b' \in B$. Then $T_1 = bTb^{-1}$ is a maximal torus in B. Since

$$g^{-1}T_1 g = (b')^{-1}\dot{w}^{-1}T\dot{w}b' = (b')^{-1}Tb' \subset B,$$

we have $T_1 \subset B \cap B'$.

10.2.10. *Corollary*. (i) *Each* $C(w)$ *is a locally closed subvariety of* G. *The boundary* $\overline{C(w)}-C(w)$ *is a union of some* $C(w')$ *with* $l(w') < l(w)$;

(ii) $C(w_0)$ *is an affine open subset of* G;

(iii) $C(w_0) \cup \coprod_{\alpha \in D} C(s_\alpha w_0)$ *is an open subset of* G. *All components of its complement have dimension* $\leq$ *dim* $G-2$.

We have already seen that $C(w)$ is locally closed. It also is irreducible, and $\dim C(w) = l(w) + \dim B$. It is easy to see that $\overline{C(w)}$ must be a union of double cosets $C(w')$, hence $\overline{C(w)}-C(w)$ is a union of such $C(w')$ with $\dim C(w') < \dim C(w)$. This implies (i). Since $C(w_0)$ is the only double coset of maximal dimension, we have (ii).

(iii) follows by noticing that the $C(s_\alpha w_0)$ ($\alpha \in D$) are the double cosets with dimension $\dim G - 1$.

The open set $C(w_0)$ is called the <u>big cell</u> defined by B, 10.2.7 also gives quite precise information about the geometric structure of the projective variety $X = G/B$. Recall (7.3.10) that we may view X as the set of all Borel subgroups. Let X_w denote the canonical image in X of $C(w)$.

10.2.11. <u>Corollary</u>. (i) <u>X_w is a locally closed subvariety of X, isomorphic to</u> $\mathbb{A}^{l(w)}$;

(ii) <u>X is the disjoint union of the</u> X_w ($w \in W$).

The X_w are called the <u>Bruhat cells</u> of X. Their closures $\overline{X}_w$ are called <u>Schubert cells</u>, they are projective algebraic varieties.

The next result, which is a consequence of 10.2.7 and the proof of 10.2.6, gives a normal form for elements of G.

10.2.12. <u>Corollary</u>. <u>Each element of G can uniquely be written in the form</u> $u\dot{w}b$, <u>with</u> $w \in W$, $u \in U_{w^{-1}}$, $b \in B$.

10.2.13. <u>Exercises</u>. Notations as before.

(1) (a) Let $w = s_1 \dots s_h$ be a shortest expression. If $s \in S$ and $l(sw) < l(w)$, there is an i such that $sw = s_1 \dots s_{i-1} s_{i+1} \dots s_h$.

(b) Denote by $w_0 \in W$ the element of maximal length, and fix a shortest expression $w_0 = s_1 \dots s_N$. For any $w \in W$ there is a subsequence $i_1 < i_2 < \dots < i_h$ of $(1,\dots,N)$ such that $w = s_{i_1} s_{i_2} \dots s_{i_h}$ is a shortest expression (Hint: if $l(w) < N$ deduce from 10.2.2 that there is $s \in S$ with $l(sw) = l(w)+1$, use induction on $N-l(w)$ and (a)).

(c) If $w = s_1 \dots s_h$ is a shortest expression, there is a shortest expression $w_0 = s_1 \dots s_h s_{h+1} \dots s_N$.

(2) Put $X = G/B$ and view it as the set of Borel groups. The group G acts on $X \times X$ by $g(xB,x'B) = ((gx)B,(gx')B)$.

(a) Let $O(w) \subset X \times X$ be the G-orbit of (B,wB). Then $w \mapsto O(w)$ defines a bijection of W onto the set of G-orbits in $X \times X$ (Hint: show that this is a reformulation of Bruhat's lemma). The Borel subgroup B',B" of G are said to be <u>in position</u> w if $(B',B'') \in O(w)$. They are <u>opposite</u> if $B' \cap B''$ is a maximal torus.

(b) B' and B" are opposite if and only if they are in position w_0 (the longest element of W).

(3) Let $C(w_0)$ be the big cell of G. Then $G = \bigcup_{w \in W} \dot{w}C(w_0)$ (Hint: pass to $X = G/B$. If $X' = \bigcup_{w \in W} \dot{w}X_{w_0}$ and $X - X' \neq \emptyset$ there would be a T-invariant point in $X - X'$, which leads to a contradiction).

(4) Let $w = s_1 \dots s_h$ be a shortest expression, with $s_i = s_{\alpha_i}$ ($\alpha_i \in D$). Denote by P_i the parabolic subgroup generated by $B, X_{\alpha_i}, X_{-\alpha_i}$ $(1 \leq i \leq h)$.

(a) $P_1 \dots P_h$ is an irreducible closed subset of G, which is a union of double cosets $C(w')$ (Hint: deduce this from

7.2.11(9) by induction on h).

Put $S_w = \{s_{i_1}\dots s_{i_m} \mid 1 \leq i_1 < \dots < i_m \leq h\}$.

(b) $P_1\dots P_h = \bigcup_{w' \in S_w} C(w')$ (Hint: use $P_i = B \cup B\dot{s}_i B$).

(c) S_w does not depend on the choice of the shortest expression. We have (the bar denoting closure) $\overline{C(w)} = \bigcup_{w' \in S_w} C(w')$, $\overline{X_w} = \bigcup_{w' \in S_w} X_{w'}$.

(d) Write $w' \leq w$ if $w' \in S_w$. Then $\leq$ defines an order relation on W (the "Bruhat-order").

(5) Assume $w, w' \in W$ and $l(w) \leq l(w')$.

(a) If $w \neq w'$ then $\overline{X_w} \cap \dot{w}_0 \overline{X_{w_0 w'}} = \phi$ (Hint: Let Y be the intersection. If $Y \neq \phi$ it contains a point $\dot{w}''B$ fixed by T. By 4(c) we have $l(w'') \leq l(w)$, $l(w_0 w'') \leq l(w_0 w')$, equality holding if and only if $w'' = w$, $w'' = w'$, respectively. Show that this leads to a contradiction).

(b) $\overline{X_w} \cap \dot{w}_0 \overline{X_{w_0 w}} = \dot{w}B$ (Hint: deduce from (a) that it suffices to prove that $X_w \cap \dot{w}_0 X_{w_0 w} = \dot{w}B$. Deduce this from 10.2.5(ii)).

(6) Let $O(w)$ be as in (2).

(a) Deduce from 4(c) that $\overline{O(w)} = \bigcup_{w' \leq w} O(w')$.

Let $w = s_1 \dots s_h$ be a shortest expression. Define

$$O(s_1, \dots, s_h) =$$

$$O\{(B_0, \dots, B_h) \in X^{h+1} \mid (B_{i-1}, B_i) \in \overline{O(s_i)}, 1 \leq i \leq h\}.$$

(b) $O(s_1, \dots, s_h)$ is a closed subvariety of X^{h+1}. The projection map $O(s_1, \dots, s_h) \to O(s_1, \dots, s_{h-1})$ is a surjective morphism, whose fibers are all isomorphic to $\mathbb{P}^1$.

(c) $(B_0,\dots,B_h) \mapsto (B_0,B_h)$ defines a morphism $\pi\colon O(s_1,\dots s_h) \to X \times X$, we have $\pi O(s_1,\dots,s_h) = \overline{O(w)}$. The restriction of π to $\pi^{-1}O(w)$ is an isomorphism.

(One can show that π is a "resolution" of $\overline{O(w)}$, see e.g. [13, no.3]).

(7) (see 7.2.11(4)). $G = \mathbb{GL}_n$. Denote by $(V_1,\dots,V_{n-1})$, $(V_1',\dots,V_{n-1}')$ two complete flags in $V = k^n$. There is a unique permutation $\sigma \in S_n$ such that there exists a basis $(e_1,\dots,e_n)$ of V with the following properties: $(e_1,\dots,e_i)$ is a basis of V_i and $(e_{\sigma 1},\dots,e_{\sigma i})$ a basis of V_i', for all i with $1 \leq i \leq n-1$. Deduce another proof of Bruhat's lemma, in this particular case (see (2) and 7.3.11(6)).

(8) Let $t \in T$. If $g = u\dot{w}b \in G$ is as in 10.2.12 and lies in the centralizer $Z_G(t)$, then $u,b,\dot{w}$ lie in $Z_G(t)$. Show that $Z_G(t)$ is generated by $Z_G(t)^0$ and the set of $\dot{w}$ with $w \in W$, $w.t = t$.

(9) Assume that X is defined over $\mathbb{F}_q$ and let F be the Frobenius morphism (3.3.15). We assume that $FB = B$. Then F also acts on $X = G/B$, which we view as the set of Borel subgroups.

(a) If $w \in W$, put $X(w) = \{B' \in X \mid B' \text{ and } FB' \text{ are in position } w\}$. Then $X(w)$ is a locally closed subset of X, on which the finite group G^F of fixed points of F operates.

(b) $G = \mathbb{GL}_n$, $B = \mathbb{T}_n$, the group of upper triangular matrices. We have $W \simeq S_n$ (10.1.12(1)). Let w be the cyclic permutation $(1,2,\dots,n)$. Using (7) show that in this case $X(w)$ is the set of flags $(V_1,\dots,V_{n-1})$ in $V = k^n$ such that there exists $v \in V$

with the property that $(v, Fv, \ldots, F^{i-1}v)$ is a basis of V_i. Define a bijection of $X(w)$ onto the complement $Y \subset \mathbb{A}^{n-1}$ of the union of all rational hyperplanes of $\mathbb{A}^{n-1}$ (i.e. hyperplanes given by an equation with coefficients in $\mathbb{F}_q$). Show that there is a bijection of $X(w)$ onto the set of $x = (x_1, \ldots, x_n)^* \in \mathbb{P}^{n-1}$ such that

$$\det\begin{pmatrix} x_1 & \cdots & x_n \\ x_1^q & \cdots & x_n^q \\ \vdots & & \\ x_1^{q^{n-1}} & \cdots & x_n^{q^{n-1}} \end{pmatrix} \neq 0.$$

10.3. <u>Parabolic subgroups</u>.

Bruhat's lemma also permits us to give an explicit description of the parabolic subgroups containing B. If I is a subset of D, denote by $W_I \subset W$ the subgroup generated by the s_α with $\alpha \in I$. Let $R_I \subset R$ be the set of roots which are linear combinations of the roots in I and put $S_I = (\bigcap_{\alpha \in I} \operatorname{Ker} \alpha)^0$, $L_I = Z_G S_I$. Then L_I is a connected reductive subgroup of G (9.3.5(i)) with maximal torus T and Borel subgroup $B \cap L_I$ 7.3.5(ii).

10.3.1. <u>Proposition</u>. <u>The root system</u> $R(L_I, T)$ <u>and the Weyl group</u> $W(L_I, T)$ <u>are</u> R_I <u>and</u> W_I, <u>respectively</u>. <u>The system of positive roots</u> $R_I^+(B \cap L_I)$ <u>defined by</u> $B \cap L_I$ <u>is</u> $R_I^+ = R_I \cap R^+$, <u>and the corresponding basis of</u> R_I <u>is</u> I.

The roots of L_I with respect to T are the non-zero weights of T in Lie L_I (9.3.7). By 4.4.7, the Lie algebra of L_I is the fixed point set of S_I in $\mathfrak{g}$ = Lie G, whence

$$\text{Lie } L_I = \mathfrak{t} + \sum_{\alpha(S_I)=1} \mathfrak{g}_\alpha .$$

Since the elements of D are linearly independent (10.1.10 (iii)) it readily follows that $\alpha(S_I) = 1$ if and only if α is a linear combination of the elements of I. It then also follows that α is an integral linear combination, with coefficients which are either all ≥ 0 or all ≤ 0, of the elements of I. This implies that $R(L_I,T) = R_I$ and $R_I^+(B \cap L_I) = R_I^+$. The simple roots in R_I^+ are those roots which cannot be written as a sum of at least two other positive roots (see the proof of 10.1.10(iii)). This implies that I is the set of simple roots in R_I^+. Then 10.1.10(i) shows that $W(L_I,T) = W_I$. This completes the proof.

10.3.2. <u>Theorem</u>. (i) $P_I = \bigcup_{w \in W_I} C(w)$ <u>is a closed subgroup of G containing</u> B;

(ii) $R_u P_I$ <u>is generated by the</u> X_α <u>with</u> $\alpha \in R^+ - R_I$;

(iii) $L_I \subset P_I$, <u>and the product morphism</u> $L_I \times R_u P_I \to P_I$ <u>is an isomorphism of varieties</u>;

(iv) <u>If</u> P <u>is any parabolic subgroup containing</u> B, <u>there is a unique</u> $I \subset S$ <u>such that</u> $P = P_I$.

It follows from 10.2.8 that P_I is a subgroup of G. Since W_I is the Weyl group of (L_I,T), it contains a unique element w_0' of maximal length. It then follows from 10.2.10(i), using the fact that $\dim C(w_0') > \dim C(w)$ for $w \in W_I$, $w \neq w_0'$, that

$C(w_0')$ is an open subset of $\overline{P_I}$ and 2.2.4(ii) shows that P_I is closed. Since $C(e) = B$ it contains B. This proves (i).
If $\alpha \in R^+ - R_I$ then $\alpha = \sum_{\beta \in D} n_\beta \beta$ with $n_\beta > 0$ for some $\beta \in D-I$.
It is then immediate that also $s_\gamma \alpha \in R^+ - R_I$ if $\gamma \in I$.
This shows that W_I stabilizes $R^+ - R_I$. Put $U_I = U \cap L_I$. It follows from 10.1.4 that the X_α with $\alpha \in R^+ - R_I$ generate a subgroup U_I' of U. We see from 10.1.2 that the product map $U_I \times U_I' \to U$ is bijective. It follows that $\bigcap_{w \in W_I} \dot{w} U \dot{w}^{-1} = (\bigcap_{w \in W_I} \dot{w} U_I \dot{w}^{-1}).U_I' = U_I' = U_I'$; for by 9.3.4 $\bigcap_{w \in W_I} \dot{w} U_I \dot{w}^{-1}$ is the unipotent radical of the reductive group L_I, hence it equals $\{e\}$. Now L_I has $L_I \cap B$ as a Borel subgroup and the Weyl group of L_I with respect to the canonical image of T is W_I: from 10.2.10(iii) one infers that the corresponding simple roots of the root system of L_I are the $\alpha \in I$. By 9.3.4 it follows that $R_u P_I = \bigcap_{w \in W_I} \dot{w} U \dot{w}^{-1} = U_I'$. This proves (ii).
If $w \in W_I$, denote by $C_G(w)$ and $C_{L_I}(w)$ double cosets in G and L_I, respectively. The bijectivity of the product map $U_I \times U_I' \to U$ then also implies that $C_G(w) = C_{L_I}(w) R_u P_I$ $(w \in W_I)$. Using Bruhat's lemma for L_I this shows that $P_I = L_I . R_u P_I$. More precisely, this argument yields that the product map $L_I \times R_u P_I \to P_I$ is bijective. Now (iii) follows by proving bijectivity of the tangent map at a suitable point (fill in the details of this argument).
If $\alpha \in D$, put $P_\alpha = P_{\{\alpha\}}$. We see from (i) that $\dim P = \dim B+1$.
Conversely, let $P \supset B$ be a connected, closed subgroup containing B with $\dim P = \dim B+1$. It follows from 9.3.8(ii) that the root system of the reductive group $P/R_u P$, with respect to

the image of T, consists of 2 elements, from which one concludes that P/R_uP has semi-simple rank 1. Hence there are only two Borel subgroups B,B' of P which contain T, which are adjacent. We conclude from 10.1.8 that $P = P_\alpha$, for some $\alpha \in D$.
Now let P be any closed subgroup containing B. It is connected (7.3.8). Let $I = \{\alpha \in D | P_\alpha \subset P\}$. We see from the preceding remark that the P_α $(\alpha \in I)$ are the closed subgroups $Q \subset P$, containing B, with dim Q = dim P+1. But the same remark, applied to the reductive group P/R_uP, shows that I is a system of simple roots for this group (relative to the image of T and B/R_uP). Bruhat's lemma, applied to P/R_uP, then shows that $P = \bigcup_{w \in W_I} C(w)$, which establishes (iv).

10.3.3. Exercises. G is a connected reductive group.
(1) Let P be a parabolic subgroup. A Levi subgroup L of P is a closed subgroup of P such that the product map $L \times R_uP \to P$ is an isomorphism of varieties. Show that if T is a maximal torus of P there is a unique Levi subgroup L containing T.
(2) Let T be a maximal torus of G and H a closed subgroup containing T. Denote by R the root system of (G,T) and put $R' = \{\alpha \in R | X_\alpha \subset H\}$. Show that H is parabolic if and only if $R' \cup (-R') = R$ (Hint: the "only if"-part follows from 10.3.2. To prove the other part, take a Borel subgroup $B' \supset T$ of H and a Borel subgroup B of G with $B \cap H = B'$, and put $R^+=R^+(B)$. If $\alpha \in R^+$ and $X_\alpha \not\subset B'$ then $X_{-\alpha} \subset H$. Using the reductive group H/R_uH this leads to the contradiction $X_\alpha \subset H \cap B = B'$. Hence B' = B).

(3) Let P and Q be two parabolic subgroups of G. Show that $(P \cap Q)R_uP$ is a parabolic subgroup (Hint: use (2)).

(4) Let $\lambda: \mathbb{G}_m \to G$ be a multiplicative 1-parameter subgroup of G. If $g \in G$, $t \in k^*$, put $\theta_{g,\lambda}(t) = \lambda(t)g\lambda(t)^{-1}$, and let $P(\lambda)$ be the set of $g \in G$ such that $\theta_{g,\lambda}$ extends to a morphism $\mathbb{A}^1 \to G$. Then $P(\lambda)$ is a closed subgroup of G (see 2.5.5(2)). Show that $P(\lambda)$ is a parabolic subgroup (Hint: take a maximal torus $T \supset S = \lambda(\mathbb{G}_m)$ and use (2)).

(5) Notations of 10.3.2. Let $I, J \subset D$. Denote by A a set of representations in W of the double cosets W_IwW_J. Then G is the disjoint union of the double cosets $P_I\dot{w}P_J$ ($w \in A$).

(6) $G = \mathbb{GL}_n$, acting in $V = k^n$. Let $F = (V_1, \ldots, V_s)$ be an arbitrary flag in V (7.2.11(4)). Take a basis $(e_1, \ldots, e_n)$ of V such that $(e_1, \ldots, e_{i_h})$ is a basis of V_h $(1 \leq h \leq s)$, let B be the Borel subgroup of G whose elements fix the flag $ke_1 \subset ke_1 + ke_2 \subset \ldots$ and denote by T the maximal torus of G whose elements fix all subspaces ke_i.

(a) The subgroup P_F of G whose elements fix F is a parabolic subgroup containing B.

(b) Let D be the basis of the root system of (G,T), defined by B. See 10.1.12(1) for an explicit description. Describe the subset I of D with $P_F = P_I$ (10.3.2).

(c) Any parabolic subgroup P of G is a P_F, for some flag F.

(d) Discuss the corresponding results for the case of special orthogonal groups (char $k \neq 2$) and symplectic groups (7.2.11(5),(6)).

Notes.

Also in this chapter most of the material is more or less standard.

In the exercises of 10.2.10 we have included some additional material, for example about Schubert varieties.

The varieties X(w) of 10.2.10(9) were introduced by Deligne and Lusztig [15], they are quite important for the representation theory of finite Chevalley groups.

11. The uniqueness theorem.

We use the notations of the preceding chapters. G is a connected reductive linear algebraic group and $\Psi = \Psi(G,T) = (X,X^\vee,R,R^\vee)$ the root datum associated to G and a maximal torus T (9.1.6). One of the main results of this chapter (11.4.2) is that Ψ determines G up to isomorphism.
In order to establish such a uniqueness result we have to study in detail how G is built up from T and the 1-parameter subgroups X_α ($\alpha \in R$). In the course of this study we shall need information about root systems of rank 2. These will be discussed first.

11.1. Root systems of rank 2.

11.1.1. We begin with the classification of root systems of rank 2. Let R be such a root system. We assume it to be contained in $V = \mathbb{R}^2$. Let $(\varepsilon_1,\varepsilon_2)$ be the canonical basis of V and (,) the symmetric bilinear form on V with $(\varepsilon_i,\varepsilon_j)=2\delta_{ij}$ $(i,j = 1,2)$. We now describe the reduced root systems (see 9.1.6). We also give a basis $D = \{\alpha,\beta\}$. W is the Weyl group.

$A_1 \times A_1$. $\alpha = \varepsilon_1$, $\beta = \varepsilon_2$,
$R = \{\pm\alpha,\pm\beta\}$

W is isomorphic to the four-group.

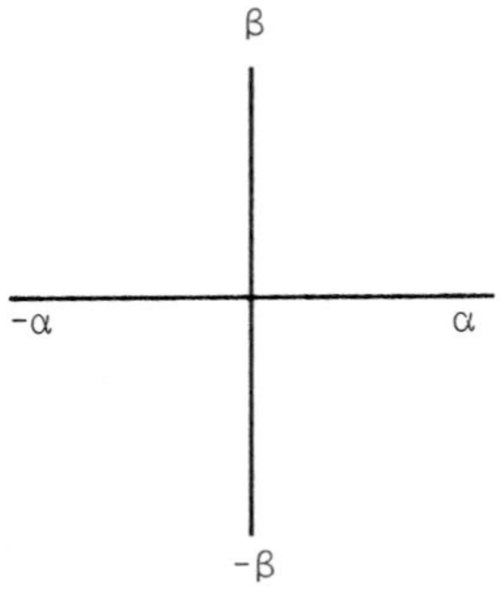

A_2. $\alpha = \varepsilon_1$, $\beta = \frac{1}{2}\varepsilon_1 + \frac{1}{2}\sqrt{3}.\varepsilon_2$,
$R = \{\pm\alpha, \pm\beta, \pm(\alpha+\beta)\}$,
W is isomorphic to the symmetric group S_3.

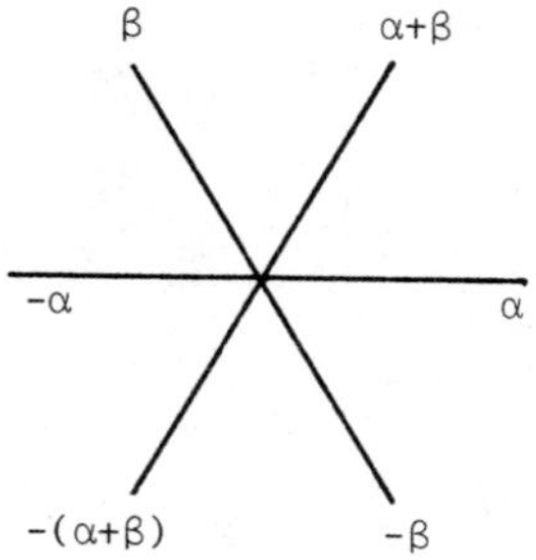

B_2. $\alpha = \varepsilon_1$, $\beta = -\varepsilon_1 + \varepsilon_2$,
$R = \{\pm\alpha, \pm\beta, \pm(\alpha+\beta), \pm(2\alpha+\beta)\}$,
W is isomorphic to the dihedral group of order 8.

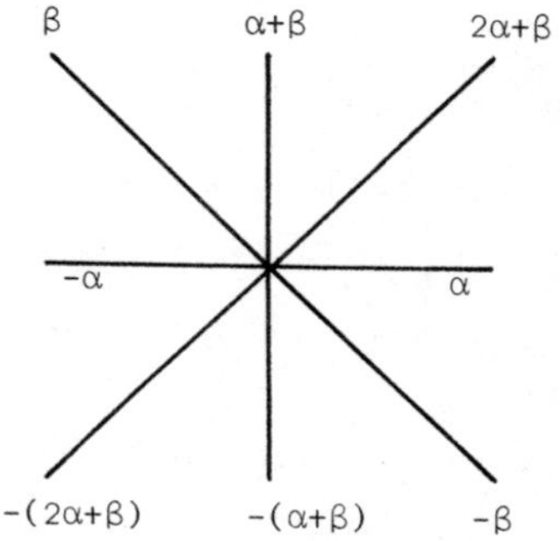

G_2. $\alpha = \varepsilon_1$, $\beta = -\frac{3}{2}\varepsilon_1 + \frac{1}{2}\sqrt{3}.\varepsilon_2$
$R = \{\pm\alpha, \pm\beta, \pm(\alpha+\beta), \pm(2\alpha+\beta), \pm(3\alpha+\beta), \pm(3\alpha+2\beta)\}$,
W is isomorphic to the dihedral group of order 12.

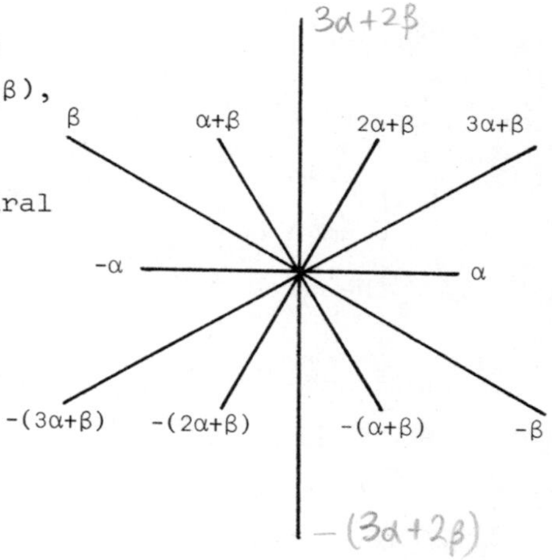

From the pictures one sees that these are indeed root systems. We may identify the coroots $\alpha^\vee$ with the vectors $2(\alpha,\alpha)^{-1}\alpha$. It is readily checked that in the cases A_2, B_2, G_2 the above symmetric bilinear form is, up to a scalar, the only W-invariant one. In the cases B_2 and G_2 there are two distinct root lengths. It is clear what is meant by long resp. short roots.

11.1.2. <u>Proposition</u>. <u>A reduced root system of rank</u> 2 <u>is isomorphic to</u> $A_1 \times A_1, A_2, B_2$ <u>or</u> G_2.
Let R be such a root system, let $D = \{\alpha,\beta\}$ be a basis of R. We know that $\langle \alpha,\beta^\vee \rangle \leq 0$, that s_α and s_β generate W and that $R = W.\alpha \cup W.\beta$ (see 10.1.10). Let $\langle \alpha,\beta^\vee \rangle\langle \beta,\alpha^\vee \rangle = a$. It follows from the facts just mentioned and 9.2.4 that there are only four possibilities for the number of elements of R, viz. 4, 6, 8, 12, corresponding to the four cases $a = 0,1,2,3$. This implies that we can only have the possibilities of 11.1.1.

11.1.3. <u>Exercise</u>. A non-reduced root system of rank 2 is isomorphic to either $\{\pm\varepsilon_1,\pm\varepsilon_2,\pm 2\varepsilon_1\}$, $\{\pm\varepsilon_1,\pm\varepsilon_2,\pm 2\varepsilon_1,\pm 2\varepsilon_2\}$ or $\{\pm\varepsilon_1,\pm\varepsilon_2,\pm\varepsilon_1\pm\varepsilon_2,\pm 2\varepsilon_1,\pm 2\varepsilon_2\}$.

We next discuss some facts, to be needed later, about general root systems, involving root systems of rank 2.
Let R be a reduced root system in the real vector space V. Denote by W the Weyl group.

11.1.4. <u>Lemma</u>. <u>Let</u> $\alpha,\beta \in R$ <u>be linearly independent roots. There is a system of positive roots</u> R^+ <u>such that</u> α <u>is a simple root of</u> R^+ <u>and</u> β <u>a positive integral linear combination of</u> α <u>and another simple root</u>.
For the notions of system of positive roots and basis see 9.2.2 and 10.1.
Let $(e_1,\ldots,e_n)$ be a basis of V with $e_1 = \alpha$, $e_2 = \beta$. Define

a total ordering on V as follows: If $x = x_1e_1+\dots x_je_j$ is a vector with $x_j \neq 0$, then $x > 0$ if and only if $x_j > 0$, and $x > y$ if $x-y > 0$. Clearly $x+y > 0$ if $x,y > 0$.
It is easy to check that $R^+ = \{\gamma \in R | \gamma > 0\}$ is a system of positive roots and that α is the smallest element of R^+ for our ordering. It follows from 10.1.10(iii) that α must be simple. It also follows that β must be a positive linear combination of simple roots which are at most β. Since these must be linear combinations of α and β, the result follows.

11.1.5. _Lemma_. _Let_ $\alpha,\beta \in R$ _be linearly independent roots_. _The set of integers_ $i \in \mathbb{Z}$ _such that_ $\beta+i\alpha \in R$ _is an interval_ $[-c,b]$ _containing_ 0. _We have_ $b-c = -\langle \alpha,\beta^\vee \rangle$, _and_ $b+c \leq 3$.
By the previous lemma we may assume that R is of rank 2, that α is a simple root and $\beta > 0$ (relative to a suitable positive system). The first statement is then easily checked in the four possible cases. To prove the last one, apply the lemma to $\beta-b\alpha$ and use 9.2.1(i).

The set of all $\beta+i\alpha \in R$ is called the α-_string_ through β. Its _length_ is $b+c$.

Denote by R^+ a system of positive roots and by D the corresponding basis. The Weyl group W is generated by the set S of reflections s_α ($\alpha \in D$). If $\alpha,\beta \in D$ denote by $R_{\alpha,\beta}$ the root system $R \cap (\mathbb{R}\alpha + \mathbb{R}\beta)$. Then $\{\alpha,\beta\}$ is a basis of $R_{\alpha,\beta}$. Its Weyl group is the subgroup $W_{\alpha,\beta}$ of W generated by s_α and s_β.

11.1.6. <u>Lemma</u>. <u>Let</u> $\alpha,\beta \in D$ <u>and</u> $w \in W$ <u>be such that</u> $w\alpha = \beta$. <u>There exist</u> $\beta_1,\ldots,\beta_{s+1} \in D$ <u>and</u> $w_1,\ldots,w_s \in W$ <u>with the following properties</u>:

(a) $w = w_s w_{s-1} \ldots w_1$,

(b) $\beta_1 = \alpha, \beta_{s+1} = \beta$ <u>and</u> $w_i\beta_i = \beta_{i+1}$ $(1 \leqslant i \leqslant s)$,

(c) <u>If</u> $\beta_i \neq \beta_{i+1}$, <u>then</u> $w_i \in W_{\beta_i,\beta_{i+1}}$ <u>and if</u> $\beta_i = \beta_{i+1}$, <u>then</u> $w_i \in W_{\beta_i,\beta_i'}$ for some $\beta_i' \in D$ $(1 \leqslant i \leqslant s)$.

This lemma expresses that the move of α into β, defined by w, can be built up from similar moves, each of them taking place in a subsystem $R_{\gamma,\delta}$ $(\gamma,\delta \in D)$.

We prove 11.1.6 by induction on the length h of w, starting with $h = 0$. Assume $h > 0$. By 10.2.2 there is $\gamma \in D$ with $w\gamma < 0$. It follows from 10.1.5(ii) that there is $w_1 \in W_{\alpha,\gamma}$ with $w_1^{-1}R^+_{\alpha,\gamma} = (w^{-1}R^+) \cap R_{\alpha,\gamma}$. If $\delta \in R^+ - R_{\alpha,\gamma}$ and $w\delta < 0$, then $ww_1^{-1}\delta < 0$, since w_1 stabilizes $R^+ - R_{\alpha,\gamma}$ (because s_α and s_γ do). Moreover $ww_1^{-1}R^+_{\alpha\gamma} \subset R^+$, but $wR^+_{\alpha\gamma} \not\subset R^+$. It follows from 10.2.2 that $l(ww_1^{-1}) < l(w) = h$. Put $\beta_1 = \alpha$, $\beta_2 = w_1\alpha$. Since $w\alpha = \beta$, we have $\alpha \in w^{-1}D \cap R_{\alpha,\gamma} = w_1^{-1}D$, hence $\beta_2 \in D \cap R_{\alpha,\gamma}$. So $\beta_2 = \beta_1$ or $\beta_2 = ww_1^{-1}\beta_1$. We can now use induction.

11.2. <u>The structure constants</u>.

Recall that for each $\alpha \in R$ we have chosen an isomorphism x_α of $\mathbb{G}_a$ onto the closed subgroup X_α of G. The choice of x_α is not unique. A first normalization of the x_α is contained in the next result.

11.2.1. Proposition. (i) *If* $a \in X_\alpha - \{1\}$ *there is a unique* $a' \in X_{-\alpha} - \{1\}$ *with* $aa'a \in N_G T$;

(ii) *The* x_α *may be chosen such that, for all* $\alpha \in R$,

$$n_\alpha = x_\alpha(1)x_{-\alpha}(-1)x_\alpha(1)$$

lies in $N_G T$ *and has image* s_α *in* $W = N_G T/T$. *We have*

(1) $x_\alpha(\xi)x_{-\alpha}(-\xi^{-1})x_\alpha(\xi) = \alpha^\vee(\xi)n_\alpha \quad (\xi \in k^*)$;

(iii) $n_\alpha^2 = t_\alpha$ *lies in* T *and* $t_\alpha = \alpha^\vee(-1)$. *Moreover,* $n_{-\alpha} = t_\alpha n_\alpha$ *and* $t_{-\alpha} = t_\alpha$;

(iv) If $(x'_\alpha)_{\alpha \in R}$ *is a second family satisfying the conditions of* (ii), *there are* $c_\alpha \in k^*$ *such that*

$$x'_\alpha(\xi) = x_\alpha(c_\alpha \xi),\ c_{-\alpha}c_\alpha = 1 \quad (\alpha \in R, \xi \in k).$$

Let $H = Z_G(\operatorname{Ker} \alpha)^0$. Then H is a connected reductive subgroup of G, and X_α, $X_{-\alpha}$ are contained in the commutator subgroup (H,H), which is isomorphic to $\mathbb{SL}_2$ or $\mathbb{PSL}_2$ (8.2.4). Since $\mathbb{PSL}_2$ is a quotient of $\mathbb{SL}_2$, it suffices to prove (i), (ii) and (iii) if $G = \mathbb{SL}_2$. We may then assume that

$$x_\alpha(\xi) = \begin{pmatrix} 1 & a\xi \\ 0 & 1 \end{pmatrix},\ x_{-\alpha}(\xi) = \begin{pmatrix} 1 & 0 \\ b\xi & 1 \end{pmatrix},\ \alpha^\vee(\eta) = \begin{pmatrix} \eta & 0 \\ 0 & \eta^{-1} \end{pmatrix},$$

if $\xi \in k$, $\eta \in k^*$, with $a,b \in k^*$. Now T is the torus $\alpha^\vee(\mathbb{G}_m)$. The proof of (i) follows by observing that if $\xi \neq 0$,

$$\begin{pmatrix} 1 & \xi \\ 0 & 1 \end{pmatrix}\begin{pmatrix} 0 & 1 \\ \eta & 1 \end{pmatrix}\begin{pmatrix} 1 & \xi \\ 0 & 1 \end{pmatrix}$$

normalizes T if and only if $\eta = -\xi^{-1}$.

We have the formulas

$$\begin{pmatrix}1 & \xi\\ 0 & 1\end{pmatrix}\begin{pmatrix}1 & 0\\ -\xi^{-1} & 1\end{pmatrix}\begin{pmatrix}1 & \xi\\ 0 & 1\end{pmatrix} = \begin{pmatrix}\xi & 0\\ 0 & \xi^{-1}\end{pmatrix}\begin{pmatrix}0 & 1\\ -1 & 0\end{pmatrix},$$

$$\begin{pmatrix}1 & 0\\ \xi & 1\end{pmatrix}\begin{pmatrix}1 & -\xi^{-1}\\ 0 & 1\end{pmatrix}\begin{pmatrix}1 & 0\\ \xi & 1\end{pmatrix} = \begin{pmatrix}\xi^{-1} & 0\\ 0 & \xi\end{pmatrix}\begin{pmatrix}0 & -1\\ 1 & 0\end{pmatrix}, \text{ if } \xi \in k^*.$$

These show that the requirement of (ii) is satisfied if and only if $ab = 1$. In that case we must take $n_\alpha = \begin{pmatrix}0 & a\\ -a^{-1} & 0\end{pmatrix}$, $n_{-\alpha} = \begin{pmatrix}0 & -a^{-1}\\ a & 0\end{pmatrix}$, $t_\alpha = \begin{pmatrix}-1 & 0\\ 0 & -1\end{pmatrix}$. The assertions (ii) and (iii) are readily checked. Finally, (iv) is a consequence of 8.2.7(iii).

Observe that if $n \in N_GT$ represents $w \in W$, we have

$$nt_\alpha n^{-1} = t_{w\alpha}.$$

11.2.2. The next result about the n_α's will be needed in chapter 12. Let B be a Borel subgroup of G containing T. Denote by R^+ the corresponding system of positive roots (9.3.2) and by D the basis of R^+. Write $B = TU$, where $U = \prod_{\gamma\in R^+} X_\gamma$. If $u = \prod_{\gamma\in R^+} x_\gamma(\xi_\gamma)$ and α is a simple root, the coordinate ξ_α is uniquely determined, i.e. independent of the choice of the order in the product. For one sees from 10.1.4 that rewriting the product in a different order only introduces elements lying in an X_γ with γ non-simple.
We write $\xi_\alpha u = \xi_\alpha$, if $\alpha \in D$.
Denote by $\Omega = TUaU$ the big cell of G (10.2.10), where $a \in N_GT$ represents the longest element of the Weyl group and put $Y = UaU$. If $x = tuau' \in \Omega$, write $\xi_\alpha x = \xi_\alpha u$ ($\alpha \in D$).
Put $Y_\alpha = \{y \in Y \mid \xi_\alpha y \neq 0\}$, this is an open subset. Let

$\Omega_\alpha = TY_\alpha$.

11.2.3. Lemma. For each simple root α, there is a morphism $\phi_\alpha : \Omega_\alpha \to \Omega_\alpha$ such that $n_\alpha(ty) = (s_\alpha t)\alpha^\vee(-\xi_\alpha y)^{-1}\phi_\alpha y$ ($t \in T$, $y \in Y_\alpha$). We have $\xi_\alpha(\phi_\alpha y) = -\xi_\alpha y$ and $\phi_\alpha^2 = \mathrm{id}$.

By 10.1.1 we have $U = X_\alpha U_1$. where $U_1 = \prod_{\substack{\gamma\in R^+\\ \gamma\neq R^+}} x_\gamma$, this is a subgroup of U. Using the formula (where $\xi \neq 0$)

$$\begin{pmatrix} 0 & 1 \\ -1 & 0 \end{pmatrix}\begin{pmatrix} 1 & \xi \\ 0 & 1 \end{pmatrix} = \begin{pmatrix} -\xi^{-1} & 0 \\ 0 & -\xi \end{pmatrix}\begin{pmatrix} 1 & -\xi \\ 0 & 1 \end{pmatrix}\begin{pmatrix} 1 & 0 \\ \xi^{-1} & 1 \end{pmatrix},$$

we find that

$$n_\alpha x_\alpha(\xi_\alpha)U_1 aU = \alpha^\vee(-\xi_\alpha)^{-1}x_\alpha(-\xi_\alpha)X_{-\alpha}U_1 aU =$$
$$\alpha^\vee(-\xi_\alpha)^{-1}x_\alpha(-\xi_\alpha)U_1 aU,$$

since $X_{-\alpha}$ normalizes U_1 (use 10.1.4) and $a^{-1}X_{-\alpha}a \subset U$. The assertions of the lemma are now readily checked.

11.2.4. We call a family $(x_\alpha)_{\alpha\in R}$ with the properties of 11.2.1(ii) a realization of R in G. Notice that, by (1), a realization determines the coroots $\alpha^\vee$.

If $\alpha,\beta \in R$ are linearly independent (i.e. $\alpha \neq \pm\beta$) we have a formula (see 10.1.4)

$$(2)\quad x_\alpha(\xi)x_\beta(\eta)x_\alpha(\xi)^{-1} = x_\beta(\eta)\prod_{\substack{i\alpha+j\beta\in R\\ i,j>0}} x_{i\alpha+j\beta}(c_{\alpha,\beta;i,j}\xi^i\eta^j),$$

the product being taken in a preassigned order.

We call the elements $c_{\alpha,\beta;i,j}$ the structure constants of G, for the given realization. If $(c'_{\alpha,\beta;i,j})$ is a set of structure constants for another realization, there are, by

11.2.1(iv), $c_\alpha \in k^*$ such that $c_\alpha c_{-\alpha} = 1$ and that

$$c'_{\alpha,\beta;i,j} = c_\alpha^{-i} c_\beta^{-j} c_{i\alpha+j\beta} c_{\alpha,\beta;i,j}.$$

We call the set $(c'_{\alpha,\beta;i,j})$ <u>equivalent</u> to $(c_{\alpha,\beta;i,j})$.
Put $c_{\alpha,\beta;0,1} = c_{\alpha,\beta;1,0} = 1$. It is readily deduced from (2) that $c_{\beta,\alpha;1,1} = -c_{\alpha,\beta;1,1}$ and that the $c_{\beta,\alpha;i,j}$ are determined by the $c_{\alpha,\beta;i,j}$.

We have to study the structure constants in detail. The first results are contained in the next proposition.
Let α and β be linearly independent roots, and denote by $(\beta-c\alpha,\ldots,\beta+b\alpha)$ the α-string through β (11.1.5).

11.2.5. <u>Proposition</u>. (i) <u>We have</u> $n_\alpha x_\beta(\xi) n^{-1} = x_{s_\alpha\beta}(d_{\alpha,\beta}\xi)$ $(\xi \in k)$, <u>where</u> $d_{\alpha,\beta} \in k^*$ <u>is given</u> by

$$d_{\alpha,\beta} = \sum_{i=\max(0,c-b)}^{c} (-1)^i c_{-\alpha,\beta;i,1} c_{\alpha,\beta-i\alpha;i+b-c,1}.$$

<u>Moreover</u>, $d_{-\alpha,\beta} = (-1)^{\langle\beta,\alpha^\vee\rangle} d_{\alpha,\beta}, d_{\alpha,\beta} d_{-\alpha,-\beta} = (-1)^{\langle\beta,\alpha^\vee\rangle}$, $d_{\alpha,\beta} d_{\alpha,-\beta} = 1$, $d_{\alpha,\beta} d_{\alpha,s_\alpha\beta} = (-1)^{\langle\beta,\alpha^\vee\rangle}$, $d_{\alpha,\alpha} = -1$;
(ii) <u>We have</u> $n_\alpha n_\beta n_\alpha^{-1} = s_\alpha(\beta)^\vee(d_{\alpha,\beta}) n_{s_\alpha\beta}$.
Here n_α is as in 11.2.1.
Assume first that $c = 0$, i.e. $\beta-\alpha \notin R$. Using (2) repeatedly it follows (check this) that
(a) $V = \prod_{\substack{i\alpha+j\beta\in R\\ i\in\mathbb{Z}, j>0}} X_{i\alpha+j\beta}$ is a subgroup of G and
$V_1 = \prod_{\substack{i\alpha+j\beta\in R\\ i\in\mathbb{Z}, j>1}} X_{i\alpha+j\beta}$ is a normal subgroup. Both V and V_1 are

normalized by X_α and $X_{-\alpha}$,

(b) $\phi(\xi_0,\ldots,\xi_b) = \prod_{i=0}^{b} x_{\beta+i\alpha}(\xi_i)V_1$ defines an isomorphism of groups $\phi\colon M = k^{b+1} \to V/V_1$.

Let H be the commutator subgroup of $Z_G(\mathrm{Ker}\ \alpha)^0$. Then X_α and $X_{-\alpha}$ generate H and it follows from the discussion in 8.2 that there is a surjective homomorphism $\psi\colon \mathbb{SL}_2 \to H$ such that

$$\psi\begin{pmatrix}1 & \xi\\ 0 & 1\end{pmatrix} = x_\alpha(\xi),\ \psi\begin{pmatrix}1 & 0\\ \xi & 1\end{pmatrix} = x_{-\alpha}(\xi)\ (\xi \in k).$$

From (a) we see that H normalizes V and V_1. Define an action ρ of $\mathbb{SL}_2$ on M by

$$\phi(\rho(g)m) = \psi(g)\phi(m)\psi(g)^{-1}V_1.$$

From (2) we see that

$$(3)\quad \begin{cases} \rho\begin{pmatrix}1 & \xi\\ 0 & 1\end{pmatrix}e_i = \sum_{h=0}^{b-i} c_{\alpha,\beta+i\alpha;h,1}\xi^h e_{i+h}, \\ \rho\begin{pmatrix}1 & 0\\ \xi & 1\end{pmatrix}e_i = \sum_{h=0}^{i} c_{-\alpha,\beta+i\alpha;h,1}\xi^h e_{i-h}, \end{cases}$$

where $(e_i)_{0\leqslant i\leqslant b}$ is the canonical basis of $M = k^{b+1}$. This shows that ρ is a rational representation of $\mathbb{SL}_2$ in M. Moreover, from

$$tx_\gamma(\xi)t^{-1} = x_\gamma(\gamma(t)\xi)\ (t \in T, \xi \in k)$$

it follows that

$$\rho\begin{pmatrix}\xi & 0\\ 0 & \xi^{-1}\end{pmatrix}e_i = \xi^{\langle \beta+i\alpha,\alpha^\vee\rangle}e_i\ (\xi \in k^*).$$

Since

$$n_\alpha^{-1}\alpha^\vee(\xi)n_\alpha = \alpha^\vee(\xi^{-1}),$$

and

$$\phi(\rho\begin{pmatrix} 0 & 1 \\ -1 & 0 \end{pmatrix}m) = n_\alpha \phi(m) n_\alpha^{-1} V_1,$$

we have

$$\rho\begin{pmatrix} \xi & 0 \\ 0 & \xi^{-1} \end{pmatrix}\rho\begin{pmatrix} 0 & 1 \\ -1 & 0 \end{pmatrix}e_i = \xi^{-\langle \beta+i\alpha,\alpha^\vee \rangle}\rho\begin{pmatrix} 0 & 1 \\ -1 & 0 \end{pmatrix}e_i.$$

Now $-\langle \beta+i\alpha,\alpha^\vee \rangle = \langle \beta+(b-i)\alpha,\alpha^\vee \rangle$ (by 11.1.5) and all integers $\langle \beta+i\alpha,\alpha^\vee \rangle$ are distinct. It follows that $\rho\begin{pmatrix} 0 & 1 \\ -1 & 0 \end{pmatrix}e_i = de_{b-i}$, with $d \in k^*$. We can determine d by using (3), comparing coefficients of e_{b-i} in either side of

$$\rho\begin{pmatrix} 1 & \xi \\ 0 & 1 \end{pmatrix}\rho\begin{pmatrix} 1 & 0 \\ -\xi^{-1} & 1 \end{pmatrix}e_i = \rho\begin{pmatrix} 0 & \xi \\ -\xi^{-1} & 0 \end{pmatrix}\rho\begin{pmatrix} 1 & -\xi \\ 0 & 1 \end{pmatrix}e_i \quad (\xi \in k^*).$$

The result is equivalent to the formula for $d_{\alpha,\beta+i\alpha}$ stated in (i). The formula for $d_{-\alpha,\beta}$ in (1) follows from $n_{-\alpha} = n_\alpha t_\alpha$ (see 11.2.1(iii)). By 11.2.1(ii) we have

$$x_{s_\alpha\beta}(d_{\alpha,\beta})x_{-s_\alpha\beta}(-d_{\alpha,-\beta})x_{s_\alpha\beta}(d_{\alpha,\beta}) =$$

$$n_\alpha x_\beta(1)x_\beta(-1)x_\beta(1)n_\alpha^{-1} = n_\alpha n_\beta n_\alpha^{-1} \in N_G T.$$

Using 11.2.1(i) we see that $d_{\alpha,-\beta} = d_{\alpha,\beta}^{-1}$ and that $n_\alpha n_\beta n_\alpha^{-1} =$ $= (s_\alpha(\beta))^\vee(d_{\alpha,\beta})n_{s_\alpha\beta}$, proving (ii). The remaining assertions of (i) also follow easily.

11.2.6. Lemma. Let $\beta,\gamma \in R$ be linearly independent and $i\beta+j\gamma \in R$, $i \geqslant 0, j \geqslant 0$. Then

$$c_{s_\alpha\beta,s_\alpha\gamma;i,j} = d_{\alpha,\beta}^{-i}d_{\alpha,\gamma}^{-i}d_{\alpha,i\beta+j\gamma}c_{\beta,\gamma;i,j},$$

the $d_{\alpha,\beta}$ being as in 11.2.5.

For any $w \in W$, we have

$$c_{\beta,\gamma;i,j}c_{-\beta,-\gamma;i,j} = c_{w\beta,w\gamma;i,j}c_{-w\beta,-w\gamma;i,j}.$$

The first formula is a consequence of (2) and the definition of the $d_{\alpha,\delta}$. The second formula follows from $d_{\alpha,\delta}d_{\alpha,-\delta} = 1$.

We next discuss the case that the root system R has rank 2. We assume that $\{\alpha,\beta\}$ is a basis as described in 11.1.1. Let $m(\alpha,\beta)$ denote the order of $s_\alpha s_\beta$.
We identify the coroot $\gamma^\vee$ with $2(\gamma,\gamma)^{-1}$.

11.2.7. Proposition. (rank R = 2). (i) The structure constants $c_{\alpha,\beta;i,j}$ $(i\alpha+j\beta \in R)$ are nonzero. The other structure constants $c_{\gamma,\delta;i,j}$ are completely determined by the $c_{\alpha,\beta;i,j}$;
(ii) If $c_{\alpha,\beta;i,j} = 1$ whenever $i\alpha+j\beta \in R$, $i,j \geqslant 0$ then all structure constants are of the form n.1, with $n \in \mathbb{Z}$. If $\gamma,\delta,\gamma+\delta \in R$, $\gamma-c\delta \in R$, $\gamma-(c+1)\delta \notin R$ then $c_{\gamma,\delta;1,1} = \pm(c+1)$.
Let $\gamma,\delta \in R$. By 11.1.4 there is $w \in W$ such that $w\gamma = \alpha$ or β and $w\delta > 0$. Since s_α and s_β generate W, it follows from 11.2.5(i) that the structure constants $c_{\gamma,\delta;i,j}$ can be determined from the $c_{\varepsilon,\zeta;h,l}$ and the $d_{\varepsilon,\zeta}$, where ε is simple and $\zeta > 0$ (and also from the $c_{\varepsilon,\zeta;h,l}$ where $\pm\,\varepsilon$ is simple and $\zeta > 0$).
Notice that $d_{\varepsilon,\zeta} = 1$ if $\varepsilon+\zeta \notin R$, $\varepsilon-\zeta \notin R$. Also recall that the $c_{\beta,\alpha;i,j}$ are determined by the $c_{\alpha,\beta;i,j}$. To prove the proposition we discuss the four possible cases for R described in 11.1.

$A_1 \times A_1$. The assertions about structure constants are vacuous.

A_2. The argument of the beginning of the proof shows that we have to deal with $c_{\alpha,\beta;1,1} = -c_{\beta,\alpha;1,1}$, $d_{\alpha,\beta}$, $d_{\beta,\alpha}$, $d_{\alpha,\alpha+\beta}$, $d_{\beta,\alpha+\beta}$. It follows from 11.2.5(i) that

$$d_{\alpha,\beta} = c_{\alpha,\beta;1,1},\ d_{\alpha,\alpha+\beta} = d_{\alpha,s_\alpha\beta} = -d^{-1}_{\alpha,\beta},$$

moreover $d_{\alpha,\beta} = -d_{\beta,\alpha}$, $d_{\beta,\alpha+\beta} = d^{-1}_{\beta,\alpha}$. The statements of (i) and (ii) readily follow.

B_2. We have to deal with $c_{\alpha,\beta;1,1}$, $c_{\alpha,\beta;2,1}$, $c_{\alpha,\alpha+\beta;1,1}$, $d_{\alpha,\beta}$, $d_{\alpha,\alpha+\beta}$, $d_{\beta,\alpha}$.
It follows from 11.2.5(i) that

$$-c_{\alpha,\beta;1,1} = c_{\beta,\alpha;1,1} = d_{\beta,\alpha} \neq 0.$$

Likewise,

$$c_{-\alpha,\beta;1,1} = d_{\beta,-\alpha} = -d^{-1}_{\beta,\alpha} \neq 0.$$

Also

$$c_{\alpha,\beta;2,1} = d_{\alpha\beta} \neq 0$$

and

$$d_{\alpha,\beta+2\alpha} = d_{\alpha,s_\alpha\beta} = d^{-1}_{\alpha,\beta}.$$

If ρ is the representation of $\mathbb{SL}_2$ used in the proof of 11.2.5 we have in the present case (where b = 2)

$$\rho\begin{pmatrix} 0 & 1 \\ -1 & 0 \end{pmatrix} e_0 = xe_2,\ \rho\begin{pmatrix} 0 & 1 \\ -1 & 0 \end{pmatrix} e_2 = x^{-1}e_0,$$

with $x = d_{\alpha,\beta}$. Since the determinant of $\rho\begin{pmatrix} 0 & 1 \\ -1 & 0 \end{pmatrix}$ must be 1, we conclude that $\rho\begin{pmatrix} 0 & 1 \\ -1 & 0 \end{pmatrix}e_1 = -e_1$, which means in the present case that $d_{\alpha,\alpha+\beta} = -1$. Hence $d_{-\alpha,\alpha+\beta} = d_{\alpha,\alpha+\beta} = -1$ (for $\langle \alpha+\beta, \alpha^\vee \rangle = 0$). By 11.2.5(i),

$$-1 = d_{-\alpha,\alpha+\beta} = 1 - c_{\alpha,\alpha+\beta;1,1}c_{-\alpha,\beta;1,1},$$

$$c_{\alpha,\alpha+\beta;1,1}c_{-\alpha,\beta;1,1} = 2.$$

The statements of (i) and (ii) follow from these relations. In fact, if $c_{\alpha,\beta;1,1} = c_{\alpha,\beta;2,1} = 1$, then $d_{\alpha,\beta} = 1$, $d_{\beta,\alpha} = -1$, $c_{\alpha,\alpha+\beta;1,1} = 2$, and $d_{\alpha,\alpha+\beta} = -1$.

G_2. This is the most complicated case. Using 11.2.5(i) we see that $-c_{\alpha,\beta;1,1} = c_{\beta,\alpha;1,1} = d_{\beta,\alpha} \neq 0$ and $c_{\alpha,\beta;3,1} = d_{\alpha,\beta} \neq 0$. It also follows that $c_{-\alpha,3\alpha+\beta;3,1} = d_{-\alpha,3\alpha+\beta} \neq 0$. We have $d_{\beta,\alpha+\beta} = d_{\beta,s_\beta\alpha} = -d_{\beta,\alpha}^{-1}$.

Next consider the α-string through β, which has length 3. We consider the corresponding rational representation ρ of $\mathbb{SL}_2$, introduced in the proof of 11.2.5. We use the notations introduced there.

The subspace M' of M generated by the elements $(\rho(g)-1)e_3$ $(g \in \mathbb{SL}_2)$ is stable under $\rho(\mathbb{SL}_2)$ and contains e_3 (because $\rho\begin{pmatrix} \xi & 0 \\ 0 & \xi^{-1} \end{pmatrix}e_3 = \xi^3 e_3$ if $\xi \in k^*$) and also $\rho\begin{pmatrix} 1 & 0 \\ 1 & 1 \end{pmatrix}e_3 - e_3 = c_{-\alpha,3\alpha+\beta;1,1}e_2 + ue_1 + ve_2$. Since, as we noticed, $c_{-\alpha,3\alpha+\beta;1,1} \neq 0$, it follows (check this) that M' contains the unique subspace ke_2 of vectors m-satisfying $\rho\begin{pmatrix} \xi & 0 \\ 0 & \xi^{-1} \end{pmatrix}m = \xi m$. Because $\rho\begin{pmatrix} 0 & 1 \\ -1 & 0 \end{pmatrix}$ interchanges ke_0, ke_3 and ke_1, ke_2, respectively, it follows that M' = M. We then conclude from

8.2.9.4(b) that ρ is the representation of $\mathbb{SL}_2$ in the space V_3^* described there. It follows (check this) that there is a unique basis (f_i) of M, with $f_i \in ke_i$ $(0 \leqslant i \leqslant 3)$ such that

$$\begin{cases} \rho\begin{pmatrix}1 & \xi\\ 0 & 1\end{pmatrix}f_0 = f_0 + \xi f_1 + \xi^2 f_2 + \xi^3 f_3, \\ \rho\begin{pmatrix}1 & \xi\\ 0 & 1\end{pmatrix}f_1 = f_1 + 2\xi f_2 + 3\xi^2 f_3 \\ \rho\begin{pmatrix}1 & \xi\\ 0 & 1\end{pmatrix}f_2 = f_2 + 3\xi f_3 \\ \rho\begin{pmatrix}1 & \xi\\ 0 & 1\end{pmatrix}f_3 = f_3. \end{cases}$$

Fixing this basis amounts to normalizing the $x_{i\alpha+\beta}$ $(0 \leqslant i \leqslant 3)$ such that $c_{\alpha,\beta;i,1} = 1$ $(0 \leqslant i \leqslant 3)$. The representation ρ is now uniquely determined. It follows that all $c_{\alpha,i\alpha+\beta}$, (and also all $c_{-\alpha,i\alpha+\beta;j,1}$) as well as all $d_{\alpha,i\alpha+\beta}$ $(0 \leqslant i \leqslant 3)$ are now fixed.

We know all structure constants $c_{\alpha,\gamma;i,j}$ and $c_{\beta,\gamma;i,j}$ $(\gamma > 0)$ except for $c_{\alpha,\beta;3,2}$, $c_{\alpha,\alpha+\beta;1,2}$, $c_{\beta,3\alpha+\beta;1,1}$ and $c_{\alpha+\beta,2\alpha+\beta;1,1}$. All $d_{\alpha,\gamma}$ and $d_{\beta,\gamma}$ $(\gamma > 0)$ are fixed and are ± 1.

To deal with these remaining structure constants we observe that it follows from formula (2) that, ϕ being as in the proof of 11.2.5, if $m,m' \in M$, $\xi \in k$ we have

$$(4) \quad \phi(m)\phi(m') = \phi(m+m')x_{3\alpha+2\beta}(f(m,m')),$$

$$(5) \quad x_\alpha(\xi)\phi(m)x_\alpha(\xi)^{-1} = \phi(\rho\begin{pmatrix}1 & \xi\\ 0 & 1\end{pmatrix}m)x_{3\alpha+2\beta}(g(\xi,m)).$$

Here $f(m,m')$ is the bilinear form on M given by

$$f(\sum_0^3 \xi_j f_j, \sum_0^3 \xi'_j f_j) = -b\xi_2\xi'_1 - a\xi_3\xi'_0,$$

with $a = c_{\beta,3\alpha+\beta;1,1}$, $b = c_{\alpha+\beta,2\alpha+\beta;\ 1,1}$.
Moreover g(,) satisfies the following "1-cocycle" condition

$$g(\xi+\eta,m) = g(\xi,\rho\begin{pmatrix}1 & \eta\\ 0 & 1\end{pmatrix}m) + g(\eta,m) \quad (\xi,\eta \in k, m \in M).$$

$g(\xi,m)$ is given explicitly by

$$g(\xi, \sum_{0}^{3} \xi_i f_i) = c\xi\xi_1^2 - b\xi^2\xi_0\xi_1 + d\xi^3\xi_0^2,$$

with $c = c_{\alpha,\alpha+\beta;1,2}$, $d = c_{\alpha,\beta;3,2}$ and b as before.
From (4) we obtain for the commutator of $\phi(m)$ and $\phi(m')$ the formula

$$(\phi(m),\phi(m')) = x_{3\alpha+2\beta}([m,m']),$$

where [,] is the alternating bilinear form on M with

$$[m,m'] = f(m,m') - f(m',m).$$

It follows from (5) that this form is invariant under all linear transformations $\rho\begin{pmatrix}1 & \xi\\ 0 & 1\end{pmatrix}$. In particular

$$0 = [f_0,f_1] = [f_0+\xi f_1+\xi^2 f_2+\xi^3 f_3, f_1+2\xi f_2+3\xi^2 f_3] =$$

$$\xi^2(3a-b),$$

whence $b = 3a$.
Inserting the expression found for g in the cocycle relation, we obtain, after a brief calculation, that $b = -c = 3d$.
Hence, if $p = \text{char } k \neq 3$, we have $b = -c = 3a$, $d = a$.
If $p = 3$ then $b = c = 0$. In that case we use the following

consequence of (4) and (5)

$$f\left(\rho\begin{pmatrix}1 & \xi\\ 0 & 1\end{pmatrix}m, \rho\begin{pmatrix}1 & \xi\\ 0 & 1\end{pmatrix}m'\right) - f(m,m') =$$

$$g(\xi,m+m') - g(\xi,m) - g(\xi,m').$$

Inserting the expressions for f and g it readily follows that we still have $d = a$.

This establishes that all structure constants are fixed, whence (i). The statements of (ii) also readily follow.

11.2.8. Proposition. (rank R = 2). *Let α and β be simple roots. Then $n_\alpha n_\beta n_\alpha \dots = n_\beta n_\alpha n_\beta \dots$ ($m(\alpha,\beta)$ factors in either side, where $m(\alpha,\beta)$ is the order of $s_\alpha s_\beta$).*

Assume $\alpha \neq \beta$, and use the notations of 11.1.1. We discuss the various cases. In the case $A_1 \times A_1$ the assertion is trivial.

A_2. By 11.2.5(ii)

$$n_\alpha n_\beta n_\alpha = (\alpha+\beta)^\vee(d_{\alpha,\beta})n_{\alpha+\beta}t_\alpha = (\alpha+\beta)^\vee(d_{\alpha,\beta})t_\beta n_{\alpha+\beta}.$$

Likewise

$$n_\beta n_\alpha n_\beta = (\alpha+\beta)^\vee(d_{\beta,\alpha})t_\alpha n_{\alpha+\beta} =$$

$$(\alpha+\beta)^\vee(d_{\alpha,\beta})(\alpha+\beta)^\vee(-1)t_\alpha n_{\alpha+\beta}.$$

In fact, we have seen in the proof of 11.2.7 that $d_{\alpha,\beta} = -d_{\beta,\alpha}$. Now $(\alpha+\beta)^\vee = \alpha^\vee + \beta^\vee$. So

$$(\alpha+\beta)^\vee(-1)t_\alpha = \alpha^\vee(-1)\beta^\vee(-1) = t_\alpha t_\beta t_\alpha = t_\beta,$$

and the assertion follows, since $m(\alpha,\beta) = 3$ in this case.

B_2. By 11.2.5(ii)

$$n_\alpha n_\beta n_\alpha = (2\alpha+\beta)^\vee(d_{\alpha,\beta})n_{2\alpha+\beta}t_\alpha = (2\alpha+\beta)^\vee(d_{\alpha,\beta})t_{\alpha+\beta}n_{2\alpha+\beta}.$$

Hence

$$(n_\alpha n_\beta)^2 = (2\alpha+\beta)^\vee(d_{\alpha,\beta})t_{\alpha+\beta}n_{2\alpha+\beta}n_\beta,$$

$$(n_\beta n_\alpha)^2 = n_\beta . n_\alpha n_\beta n_\alpha = n_\beta(2\alpha+\beta)^\vee(d_{\alpha,\beta})t_{\alpha+\beta}n_\beta^{-1}n_{2\alpha+\beta}n_\beta =$$

$$(2\alpha+\beta)^\vee(d_{\alpha,\beta})t_\alpha n_{2\alpha+\beta}n_\beta,$$

since n_β and $n_{2\alpha+\beta}$ commute, whereas $s_\beta(2\alpha+\beta) = \alpha$.
In this case, because $(\alpha+\beta)^\vee = \alpha^\vee + 2\beta^\vee$,

$$t_{\alpha+\beta} = (\alpha+\beta)^\vee(-1) = \alpha^\vee(-1) = t_\alpha,$$

and the assertion follows since $m(\alpha,\beta) = 4$.

G_2. In this case

$$n_\alpha n_\beta n_\alpha = (3\alpha+\beta)^\vee(d_{\alpha,\beta})t_{2\alpha+\beta}n_{3\alpha+\beta},$$

$$n_\beta n_\alpha n_\beta = (\alpha+\beta)^\vee(d_{\beta,\alpha})t_{3\alpha+2\beta}n_{\alpha+\beta}.$$

As $m(\alpha,\beta) = 6$, we have to show that $n_\alpha n_\beta n_\alpha$ and $n_\beta n_\alpha n_\beta$ commute. Since $n_{\beta+\alpha}$ and $n_{\beta+3\alpha}$ commute, it suffices to prove that

$$(3\alpha+\beta)^\vee(d_{\alpha,\beta})t_{2\alpha+\beta}n_{3\alpha+\beta}((\alpha+\beta)^\vee(d_{\beta,\alpha})t_{3\alpha+2\beta})n_{3\alpha+\beta}^{-1} =$$

$$= (\alpha+\beta)^\vee(d_{\beta,\alpha})t_{2\alpha+\beta}n_{3\alpha+\beta}((3\alpha+\beta)^\vee(d_{\alpha,\beta})t_{2\alpha+\beta})n_{\alpha+\beta}^{-1}.$$

Now

$$n_{3\alpha+\beta}t_{3\alpha+2\beta}n_{3\alpha+\beta}^{-1} = t_\beta,\ n_{\alpha+\beta}t_{2\alpha+\beta}n_{\alpha+\beta}^{-1} = t_\alpha,$$

$$n_{3\alpha+\beta}(\alpha+\beta)^\vee(\xi)n_{3\alpha+\beta}^{-1} = (\alpha+\beta)^\vee(\xi).$$

The formula to be proved then simplifies to

$$t_{2\alpha+\beta}t_\beta = t_{3\alpha+2\beta}t_\alpha.$$

This is a consequence of

$$(2\alpha+\beta)^\vee+\beta^\vee = 2\alpha^\vee+4\beta^\vee,\ (3\alpha+2\beta)^\vee+\alpha^\vee = 2\alpha^\vee+2\beta^\vee.$$

We have thus checked the four rank 2 cases, and 11.2.8 is proved.

11.2.9. Now assume again R arbitrary, not necessarily of rank 2. We keep the previous notations.
Assume $w \in W$, and let $w = s_{\alpha_1}\dots s_{\alpha_h}$ be a shortest expression, the α_i being simple roots. It follows from 11.2.8 and 10.2.3 that $\phi(w) = n_{\alpha_1}n_{\alpha_2}\dots n_{\alpha_h} \in N_GT$ depends only on w, and not on the choice of the shortest expression. $\phi(w)$ is a representative of $w \in W$ in N_GT, we have $\phi(s_\alpha) = n_\alpha$ $(\alpha \in D)$.

11.2.10. <u>Exercise</u>. (a) Let $\phi(w)$ be as in 11.2.9. If $w,w' \in W$ then $\phi(ww') = \phi(w)\phi(w')c(w,w')$, with $c(w,w') \in T$, $c(w,w')^2=1$. We have $c(w,w') = 1$ if $l(w,w') = l(w) + l(w')$.
(b) Let $\overline{W}$ be the group generated by the n_α $(\alpha \in D)$. Then $\overline{W}$ is an extension of W by an elementary abelian 2-group. (More details about $\overline{W}$ can be found in [35]).

11.2.11. <u>Proposition</u>. <u>If</u> $\gamma,\delta \in D$ <u>and</u> $w \in W$ <u>are such that</u> $w\gamma = \delta$ <u>then</u> $\phi(w)x_\gamma(\xi)\phi(w)^{-1} = x_\delta(\xi)$ $(\xi \in k)$.

We may assume that $w \neq 1$. We deduce from 11.1.6 that it suffices to prove 11.2.11 in the case that R has rank 2. We discuss again the four possible cases. α and β are as in 11.1.1. In case $A_1 \times A_1$ there is nothing to prove.

A_2. If $w \neq 1$ then $w\alpha \neq \alpha$, $w\beta \neq \beta$. By symmetry, it suffices to consider the case $\gamma = \alpha$, $\delta = \beta$. We then must have $w = s_\alpha s_\beta$. Hence $\phi(w)x_\alpha(\xi)\phi(w)^{-1} = x_\beta(a\xi)$, with $a = d_{\alpha,\alpha+\beta}d_{\beta,\alpha}$. By what we established in the proof of 11.2.7 it follows that $a = 1$.

B_2. Since α and β have different lengths we have $\gamma = \delta$. If $w\beta = \beta$ then $w = s_{2\alpha+\beta}$ and the result follows because $n_{2\alpha+\beta}$ and n_β commute.
If $w\alpha = \alpha$ then $w = s_{\alpha+\beta} = s_\beta s_\alpha s_\beta$. Using that $d_{\alpha,\alpha+\beta} = -1$ (which was established in the proof of 11.2.7) we see that $\phi(w)x_\alpha(\xi)\phi(w)^{-1} = x_\alpha(a\xi)$, with $a = -d_{\beta,\alpha+\beta}d_{\beta,\alpha} = -d_{\beta,s_\beta\alpha}d_{\beta,\alpha}$. By 11.2.5(i) it follows that $a = -(-1)^{\langle \alpha,\beta^\vee \rangle} = 1$.

G_2. As in the previous case, we are reduced to considering the situation that $w\beta = \beta$. Then $w = s_{2\alpha+\beta} = n_\alpha n_\beta n_\alpha n_\beta n_\alpha$. Put $\phi(w)x_\beta(\xi)\phi(w)^{-1} = x_\beta(a\xi)$. It follows from 11.2.1 that $\phi(w)x_{-\beta}(\xi)\phi(w)^{-1} = x_{-\beta}(-a^{-1}\xi)$. Hence

$$\phi(w)n_\beta\phi(w)^{-1} = \phi(w)x_\beta(1)x_{-\beta}(-1)x_\beta(1)\phi(w)^{-1} =$$

$$x_\beta(a)x_{-\beta}(-a^{-1})x_\beta(a) = \beta^\vee(a)n_\beta.$$

This shows that $(n_\alpha n_\beta)^3 = \phi(w)n_\beta = \beta^\vee(a)n_\beta\phi(w) = \beta^\vee(a)(n_\beta n_\alpha)^3$. By 11.2.8 we conclude that $\beta^\vee(a) = 1$ and $a = \alpha(\beta^\vee(a)) = 1$.

11.3. G as an abstract group.

11.3.1. Let $\Psi = (X, X^{\vee}, R, R^{\vee})$ be the root datum of the connected reductive linear algebraic group G. We shall show, eventually, that Ψ completely determines G. As a preparation, we discuss now a presentation of G as an abstract group.
Let R^+ be a system of positive roots in R and D the corresponding basis. If $\alpha, \beta \in D$ denote by $R_{\alpha,\beta}$ the root system $(\mathbb{R}\alpha + \mathbb{R}\beta) \cap R$. If $\alpha \neq \beta$ it has rank 2, and $\{\alpha, \beta\}$ is a basis.
Put $R_1 = \bigcup_{\alpha,\beta \in D} R_{\alpha,\beta}$.
Fix a realization $(x_\alpha)_{\alpha \in R}$ of R in G (11.2.4) and let the $c_{\alpha,\beta;i,j}$ be corresponding structure constants.
We shall now give a presentation of a group $\tilde{G}$. This will only involve our field k, Ψ and structure constants $c_{\gamma,\delta;i,j}$ for two roots γ, δ which are both contained in some $R_{\alpha,\beta}$ $(\alpha, \beta \in D)$. The ingredients of the presentation are the following.

(a) $\tilde{T} = \mathrm{Hom}(X, k^*)$, the group of homomorphisms of abelian groups. Define a homomorphism $\pi: \tilde{T} \to T$ by $x(\pi\tilde{t}) = \tilde{t}(x)$ $(x \in X)$. It follows from the discussion of tori in 2.5 that π is an isomorphic (check this, e.g. by taking $T = (k^*)^n$). The inverse of π is given by $\pi^{-1}t(x) = x(t)$ $(t \in T, x \in X)$. If $x \in X$ define a homomorphism $\tilde{x}: \tilde{T} \to k^*$ by $\tilde{x}(\tilde{t}) = \tilde{t}(x)$, if $u \in X^{\vee}$ define a homomorphism $\tilde{u}: k^* \to \tilde{T}$ by $\tilde{u}(\xi)(x) = \xi^{\langle x,u \rangle}$ $(\xi \in k^*, x \in X)$. The Weyl group W acts on $\tilde{T}$ by $(w\tilde{t})x = \tilde{t}(w^{-1}x)$ $(w \in W, x \in X, \tilde{t} \in \tilde{T})$.

(b) If $\gamma \in R_1$, $\xi \in k$ we have a generator $\tilde{x}_\gamma(\xi)$.
We impose the relations (where $\xi, \eta \in k$)

(1) $\tilde{x}_\gamma(\xi+\eta) = \tilde{x}_\gamma(\xi)\tilde{x}_\gamma(\eta)$,

(2) $\tilde{x}_\gamma(\xi)\tilde{x}_\delta(\eta)\tilde{x}_\gamma(\xi)^{-1} = \tilde{x}_\delta(\eta) \prod_{\substack{i\gamma+j\delta\in R \\ i,j>0}} \tilde{x}_{i\gamma+j\delta}(c_{\gamma,\delta;i,j}\xi^i\eta^j)$,

if $\gamma,\delta \in R_1$ are linearly independent and both contained in some system $R_{\alpha,\beta}$ ($\alpha,\beta \in D$).

We also impose the relations

(3) $\tilde{t}\tilde{x}_\gamma(\xi)\tilde{t}^{-1} = \tilde{x}_\gamma(\tilde{\gamma}(\tilde{t})\xi)$,

if $\tilde{t} \in \tilde{T}$, $\gamma \in R_1$, $\xi \in k$.

(c) If $\gamma \in R_1$ put

$$\tilde{n}_\gamma = \tilde{x}_\gamma(1)\tilde{x}_{-\gamma}(-1)\tilde{x}_\gamma(1),$$

then $\tilde{n}_\gamma$ acts on $\tilde{T}$ via

$$\tilde{n}_\gamma\tilde{t}\tilde{n}_\gamma^{-1}(x) = \tilde{t}(s_\gamma x).$$

We require

(4) $\tilde{n}_\gamma\tilde{x}_\gamma(\xi)\tilde{n}_\gamma^{-1} = \tilde{x}_{-\gamma}(-\xi)$ ($\xi \in k$),

(5) $\tilde{n}_\gamma^{\,2} = \tilde{t}_\gamma \in \tilde{T}$ with $\tilde{t}_\gamma(x) = (-1)^{\langle x,\gamma^\vee\rangle}$ ($x \in X$),

(6) $\tilde{n}_\alpha\tilde{n}_\beta\tilde{n}_\alpha\ldots = \tilde{n}_\beta\tilde{n}_\alpha\tilde{n}_\beta,\ldots$ ($m(\alpha,\beta)$ factors in either side)

if $\alpha,\beta \in D$ and $m(\alpha,\beta)$ is the order of $s_\alpha s_\beta$ ($\alpha \neq \beta$),

(7) $\tilde{x}_\gamma(\xi)\tilde{x}_{-\gamma}(-\xi^{-1})\tilde{x}_\gamma(\xi) = \tilde{\gamma}^\vee(\xi)n_\gamma$, if $\gamma \in R_1$, $\xi \in k^*$.

Let $\tilde{G}$ be the group generated by $\tilde{T}$ and the $\tilde{x}_\gamma(\xi)$ ($\gamma \in R_1$, $\xi \in k$), subject to the relations (1),...,(7). Since the corresponding relations hold in G, it follows that the homomorphism π: $\tilde{T} \to T$ can be extended to a homomorphism π: $\tilde{G} \to G$.

11.3.2. <u>Theorem</u>. π <u>is an isomorphism of abstract groups</u> $\tilde{G} \to G$.

It should be mentioned that the relations we have imposed

are not independent, and that some of them could be omitted (see 11.3.3(2)).

First observe that (compare 11.2.9) we may, using (6), define a map $\tilde{\phi}$: $W \to \tilde{G}$ by $\phi\tilde{w} = \tilde{n}_{\alpha_1} \ldots \tilde{n}_{\alpha_h}$, if $w = s_{\alpha_1} \ldots s_{\alpha_h}$ is a shortest expression.

If $\gamma \in R_1$, put $\tilde{X}_\gamma = \tilde{x}_\gamma(k)$, this is a subgroup of $\tilde{G}$ which is normalized by $\tilde{T}$. Let γ and δ be two linearly independent roots, contained in some subsystem $R_{\alpha,\beta}$ ($\alpha,\beta \in D$). Let $\tilde{U}_{\gamma,\delta}$ be the subgroup of $\tilde{G}$ generated by the $\tilde{x}_{i\gamma+j\delta}$ ($i+j \geqslant 1$). Prescribe some order on the roots $i\gamma+j\delta$ in R, with $i \geqslant 0$, $j \geqslant 0$. It follows from (2) by recurrence that

$$\tilde{U}_{\gamma,\delta,n} = \prod_{\substack{i\gamma+j\delta\in R \\ i+j \geqslant n}} \tilde{X}_{i\gamma+j\delta}$$

is a descending set of normal subgroups of $\tilde{U}_{\gamma,\delta}$, such that $\tilde{U}_{\gamma,\delta,n}/\tilde{U}_{\gamma,\delta,n+1}$ is commutative. It also follows that $\tilde{U}_{\gamma,\delta} = \tilde{U}_{\gamma,\delta,1}$. Put $U_{\gamma,\delta} = \pi\tilde{U}_{\gamma,\delta}$, this is the subgroup of G generated by X_γ and X_δ. But it follows (see 10.1.1) that the product map

$$\prod_{\substack{i\gamma+j\delta\in R \\ i+j \geqslant 1}} X_{i\gamma+j\delta} \to U_{\gamma,\delta}$$

is bijective. From this one conclude that π defines an <u>isomorphism</u> $\tilde{U}_{\gamma,\delta} \to U_{\gamma,\delta}$. In particular, if $\alpha,\beta \in D$, the product map $\prod_{\delta\in R^+_{\alpha,\beta}} X_\delta \to U_{\alpha,\beta}$ is bijective (for any prescribed ordering of $R^+_{\alpha,\beta}$).

Now let γ and δ be linearly independent in some $R_{\alpha,\beta}$, with γ simple. Put

$$\tilde{V} = \prod_{\substack{j>0 \\ i \in \mathbb{Z}}} X_{i\gamma+j\delta}.$$

If $\delta > 0$ we have $\tilde{V} \subset \tilde{U}_{\alpha,\beta}$ and if $\delta < 0$ we have $\tilde{V} \subset \prod_{\varepsilon \in R^+_{\alpha,\beta}} X_{-\varepsilon} = \tilde{U}^-_{\alpha,\beta}$. It follows (check this) that $\tilde{V}$ is a subgroup of $\tilde{G}$ which is mapped isomorphically by π onto

$$V = \prod_{\substack{j>0 \\ i \in \mathbb{Z}}} X_{i\gamma+j\delta}$$

By (2), both $\tilde{X}_\gamma$ and $\tilde{X}_{-\gamma}$ normalize $\tilde{V}$, hence so does $\tilde{n}_\gamma$. So $\tilde{n}_\gamma \tilde{X}_\delta \tilde{n}_\gamma^{-1} \subset \tilde{V}$. Since $\pi(\tilde{n}_\gamma \tilde{X}_\delta \tilde{n}_\gamma^{-1}) = n_\gamma X_\delta n_\gamma^{-1} = X_{s_\gamma \delta}$, we must have $\tilde{n}_\gamma \tilde{X}_\delta \tilde{n}_\gamma^{-1} = \tilde{X}_{s_\gamma \delta}$.

It also follows that if $w \in W_{\alpha,\beta}$, the Weyl group of $R_{\alpha,\beta}$, and $\gamma \in R_{\alpha,\beta}$, then $\tilde{\phi}(w)\tilde{X}_\gamma \phi(w)^{-1} = \tilde{X}_{w\gamma}$.

We can now define $\tilde{X}_\gamma$ for any $\gamma \in R$. If $\gamma \in R$, there is $w \in W$ and $\alpha \in D$ such that $\gamma = w\alpha$ (10.1.10(ii)). Put $\tilde{X}_\gamma = \tilde{\phi}(w)\tilde{X}_\alpha \tilde{\phi}(w)^{-1}$. Using what we just established, together with 11.1.6, we see that this definition is independent of the choice of w and α. It also follows that $\tilde{X}_{w\gamma} = \tilde{\phi}(w)\tilde{X}_\gamma \phi(w)^{-1}$. Also, (3) holds for all $\gamma \in R$. Now let γ and δ be independent roots in R. By 11.1.4 there exists $w \in W$ and $\alpha, \beta \in D$ such that $w\gamma, w\delta \in R_{\alpha,\beta}$. Hence we have

$$(\tilde{X}_\gamma, \tilde{X}_\delta) = \tilde{\phi}(w)\,(\tilde{X}_{w\gamma}, \tilde{X}_{w\delta})\,\tilde{\phi}(w)^{-1},$$

and (2) implies that

$$(\tilde{X}_\gamma, \tilde{X}_\delta) \subset \prod_{\substack{i,j \geq 1 \\ i\gamma+j\delta \in R}} \tilde{X}_{i\gamma+j\delta}.$$

If $\gamma \in R$ and $\gamma = \sum_{\alpha \in D} n_\alpha \alpha$, we call $\mathrm{ht}\gamma = \sum_{\alpha \in D} n_\alpha$ the <u>height</u> of α.

By recurrence on n one proves that

$$\widetilde{U}_n = \prod_{ht\gamma \geq n} \widetilde{X}_\gamma$$

gives a descending set of normal subgroups of the group $\widetilde{U}$ generated by the $\widetilde{X}_\gamma$ with $\gamma > 0$, such that $\widetilde{U}_n/\widetilde{U}_{n+1}$ is commutative. An argument similar to the one given in the beginning of this proof now shows that π induces an isomorphism of $\widetilde{U}$ onto the subgroup U of G generated by the X_γ with $\gamma > 0$, and also an isomorphism of the group $\widetilde{B}$ generated by $\widetilde{T}$ and $\widetilde{U}$ onto $B = T.U$.

If $w \in W$, let $\widetilde{U}_w \subset \widetilde{U}$ be the subgroup of $\widetilde{U}$ generated by the $\alpha \in R$ with $w\alpha < 0$ and let $\widetilde{C}(w) = \widetilde{B}\widetilde{\phi}(w)\widetilde{B}$. The proof of 10.2.6 can be adapted to show that the product map $(x,y) \mapsto x\widetilde{\phi}(w)y$ defines a bijection $\widetilde{U}_w \times \widetilde{B} \to \widetilde{C}(w)$. The argument of the proof of 10.2.7 can also be adapted and shows that Bruhat's lemma holds in this case: G is the disjoint union of the sets $\widetilde{C}(w)$, $w \in W$ (check this, observe that here one has to use relation (7)). But π defines a bijection $\widetilde{C}(w) \to C(w)$ $(= B\dot{w}B)$. This proves the theorem.

11.3.3. Exercises. (1) Let F be any field. Let Γ be the group generated by symbols $x(\xi), y(\eta)$ $(\xi,\eta \in F)$, subject to the following relations:

(A) $x(\xi+\eta) = x(\xi)x(\eta)$, $y(\xi+\eta) = y(\xi)y(\eta)$ $(\xi,\eta \in F)$.

Put $w(\xi) = x(\xi)y(-\xi^{-1})x(\xi)$ $(\xi \in F^*)$. We also impose the relations

(B) $w(\xi)x(\eta)w(\xi)^{-1} = y(-\xi^2\eta)$ $(\xi \in F^*, \eta \in F)$.

Let $a(\xi) = w(\xi)w(1)^{-1}$ $(\xi \in F^*)$.

(a) $a(\xi)x(\eta)a(\xi)^{-1} = x(\xi^2\eta)$, $a(\xi)y(\eta)a(\xi)^{-1} = y(\xi^{-2}\eta)$ $(\xi \in F^*, \eta \in F)$.

(b) Let X be the subgroup of Γ of the $x(\xi)$ $(\xi \in F)$ and A the subgroup generated by the $a(\xi)$ $(\xi \in F^*)$. Establish "Bruhat's lemma" for G: $\Gamma = XA \cup Xw(1)XA$.

(c) There is an obvious homomorphism $\pi: \Gamma \to SL_2(F)$. Show that Ker π is a central subgroup of Γ, which is contained in A.

(d) Let Γ' be the group generated by symbols $x'(\xi), y'(\eta)$ $(\xi, \eta \in F)$, subject to relations (A'), (B') as before, and moreover also

(C') $a'(\xi\eta) = a'(\xi)a'(\eta)$ $(\xi, \eta \in F)$.

There is an obvious homomorphism $\pi': \Gamma' \to SL_2(F)$. Show that π' is an isomorphism.

(2) Let G be a connected semi-simple linear algebraic group over k, with root system R (relative to a maximal torus T). Let $(x_\alpha)_{\alpha \in R}$ be a realization of R in G, with structure constants $(c_{\alpha,\beta;i,j})$, for some order on R. Denote by Γ the group generated by symbols $\overline{x}_\alpha(\xi)$ $(\alpha \in R, \xi \in k)$, subject to the relations

(A) $\overline{x}_\alpha(\xi+\eta) = \overline{x}_\alpha(\xi)\overline{x}_\alpha(\eta)$ $(\alpha \in R,\ \xi, \eta \in k)$,

(B) $\overline{x}_\alpha(\xi)\overline{x}_\beta(\eta)\overline{x}_\alpha(\xi)^{-1} = \overline{x}_\beta(\eta) \prod_{\substack{i\alpha+j\beta \in R \\ i,j > 0}} \overline{x}_{i\alpha+j\beta}(c_{\alpha,\beta;i,j}\xi^i\eta^j)$,

if $\alpha, \beta \in R$ are linearly independent. Denote by $\pi: \Gamma \to G$ the obvious homomorphism.

(a) Let R^+ be a system of positive roots and denote by $\overline{U}$ the subgroup of Γ generated by the $\overline{x}_\alpha(\xi)$ with $\alpha \in R^+$. Then the restriction of π to $\overline{U}$ is bijective and any element of $\overline{U}$ can uniquely be written in the form $\prod_{\alpha \in R^+} \overline{x}_\alpha(\xi_\alpha)$.

(b) Put $\bar{n}_\alpha(\xi) = \bar{x}_\alpha(\xi)\bar{x}_{-\alpha}(-\xi^{-1})\bar{x}_\alpha(\xi)$ ($\alpha \in R, \xi \in k^*$). If α and β are linearly independent roots then

$$\bar{n}_\alpha(\xi)\bar{x}_\beta(\xi)\bar{n}_\alpha(\xi)^{-1} = \bar{x}_{s_\alpha\beta}(d_{\alpha,\beta}\xi^{\langle\beta,\alpha^\vee\rangle}\eta),$$

where $d_{\alpha,\beta}$ is as in 11.2.5 (Hint: adapt the proof of 11.2.5, and use (a)).

(c) Let $\alpha \in R$ and assume that there is $\beta \in R$, linearly independent of α such that $\langle\alpha,\beta^\vee\rangle \neq 0$. Then

$$\bar{n}_\alpha(\xi)\bar{x}_\alpha(\eta)\bar{n}_\alpha(\xi)^{-1} = \bar{x}_{-\alpha}(-\xi^2\eta) \quad (\xi \in k^*, \eta \in k).$$

(Hint: We may assume that rank R = 2 and that there is a basis $\{\gamma,\delta\}$ with $\gamma \neq \alpha$, $\delta \neq \alpha$. Using 11.2.7(i) we find a relation

$$(\bar{x}_\gamma(\rho),\bar{x}_\delta(\sigma)) = \bar{x}_\alpha(\eta) \prod_{\substack{i,j>0\\ i\gamma+j\delta\neq\alpha}} \bar{x}_{i\gamma+j\delta}(\eta_{i,j}).$$

Conjugate both sides with $\bar{n}_\alpha(\xi)$. The result follows from (a), (b) and the fact that the analogue of assertion relation holds in G).

(d) R is irreducible if it is not the union of two mutually orthogonal proper subsystems (see 9.4.3). If R is irreducible, relation (4) of 11.3.1 is a consequence of (1) and (2).

The group Γ of this exercise can be defined with an arbitrary underlying field. For details and more properties we refer to [32] or [34].

(e) Establish an analogue of Bruhat's lemma for Γ.

11.3.4. As a first consequence of 11.3.2 we shall deduce a uniqueness statement about the structure constants.
Before doing this we recall the definition of the Dynkin diagram of the root system R, defined by the basis D. This is a graph $\mathcal{D}$ whose vertices are the elements of D. The distinct vertices α,β are joined by $\langle\alpha,\beta^\vee\rangle\langle\beta,\alpha^\vee\rangle$ bonds, with an arrow pointing towards the shorter root, if α and β have different lengths. α and β are not joined if and only if they are orthogonal, with respect to a W-invariant positive definite symmetric bilinear form on V. We recall a number of properties. For proofs we refer to the literature (e.g. [8,Ch.VI]).

(a) R is said to be <u>irreducible</u> if it is not the union of two mutually orthogonal proper subsystems. An arbitrary root system R is the union of mutually orthogonal irreducible ones, these are uniquely determined (see also 9.4.3).
Assume R to be irreducible.
(b) $\mathcal{D}$ is connected and is a tree, i.e. contains no circuits.
(c) If a triple bond occurs then R is of type G_2.
(d) If a double bond occurs then $\mathcal{D}$ is a chain, and only one such bond occurs. The set $\{(\alpha,\alpha)|\alpha \in D\}$ contains 2 elements.
The possible Dynkin diagrams of irreducible root systems are listed in [8, Ch. VI, p. 197].
We say D or R is <u>simply laced</u> if no multiple bonds occur in $\mathcal{D}$. In the situation of (d), one has <u>long</u> and <u>short</u> roots in R.

11.3.5. <u>Lemma. There is a decomposition into two disjoint subsets</u> $D = D_1 \cup D_2$ <u>such that any two distinct roots in</u> D_i

(i=1,2) are orthogonal.

We may assume R to be irrducible. Let $\alpha \in \mathcal{D}$ be an endpoint, i.e. such that it is joined to only one other vertex β. Such an α exists because $\mathcal{D}$ contains no circuits. Then $D = D-\{\alpha\}$ is a basis of a root system of smaller rank, and we may assume the lemma to be true for D'. Hence we have a decomposition $D' = D_1' \cup D_2'$ with the required properties. If $\beta \in D_1'$, take $D_1 = D_1'$, $D_2 = D_2' \cup \{\alpha\}$.

Now let R and D be as before, in 11.3.1. Assume a decomposition $D = D_1 \cup D_2$ with the properties of 11.3.5 is given.

11.3.6. Theorem. (i) There exist unique structure constants $(c_{\gamma,\delta;i,j})$ (for a suitable realization of G) with the following properties:

(a) If $\alpha \in D_1$, $\beta \in D_2$ have the same length and $\alpha+\beta \in R$, then $c_{\alpha,\beta;1,1} = 1$,

(b) If $\alpha \in D$ is a short root and $\beta \in D$ a long one, with $\alpha+\beta \in R$, then all structure constants $c_{\alpha,\beta;i,j}$ $(i\alpha+j\beta \in R,$ $i,j > 0)$ equal 1;

(ii) There exists a realization $(x_\gamma)_{\gamma \in R}$ such that the corresponding structure constants satisfy the conditions of (i) and such that, moreover

$$\phi(w)x_\gamma(\xi)\phi(w)^{-1} = x_{w\gamma}(\pm\, \xi),$$

if $\gamma \in R$, $w \in W$, $\xi \in k$ (ϕ being as in 11.2.9).

In that case all structure constants are of the form n.1, with $n \in \mathbb{Z}$. If $\gamma,\delta \in R$, $\gamma+\delta \in R$, $\gamma-c\delta \in R$, $\gamma-(c+1)\delta \notin R$

then $c_{\gamma,\delta;1,1} = \pm(c+1)$ *and* $c_{\gamma,\delta;1,1}c_{-\gamma,-\delta;1,1} = -(c+1)^2$.

(iii) *If* $(x'_\gamma)_{\gamma\in R}$ *is another realization with the properties of* (ii) *there exist signs* $\varepsilon_\gamma = \pm 1$ $(\gamma \in R)$, *with* $\varepsilon_\gamma = \varepsilon_{-\gamma}$, *such that* $x'_\gamma(\xi) = x_\gamma(\varepsilon_\gamma \xi)$ $(\gamma \in R, \xi \in k)$.

It follows from 11.2.7(i) that we can find a realization of R such that the corresponding structure constants satisfy (a) and (b). It also follows from 11.2.7(i) that all $c_{\gamma,\delta;i,j}$, with $\gamma,\delta \in R_{\alpha,\beta}$ for some pair $\alpha,\beta \in D$, are fixed. But 11.3.2 shows that the root datum Ψ, together with these $c_{\gamma,\delta;i,j}$, completely determines the group structure of G. This implies that all structure constants are fixed. This establishes (i).

The analysis of the rank 2 cases made in the proof of 11.2.7 shows that if rank R = 2 there exists a realization with the required properties. Now assume R arbitrary and assume that for all γ in some $R_{\alpha,\beta}$ $(\alpha,\beta \in D)$ the x_γ have already been defined, so as to satisfy the formula of (ii).

Assume $\alpha,\alpha' \in D$, $w,w' \in W$, $w\alpha = w'\alpha'$. It then follows from 11.2.11 that

$$\phi(w)x_\alpha(\xi)\phi(w)^{-1} = \phi(w')x_{\alpha'}(\varepsilon\xi)\phi(w')^{-1},$$

with $\varepsilon = \alpha'(c(w,w^{-1}w'))$, where c(,) is as in 11.2.10. It follows from 11.2.10 that $\varepsilon = \pm 1$. We conclude (check this) that a realization (x_γ) with the first property of (ii) can be found as follows: If $\gamma \in R$, $\gamma = w\alpha$ with $w \in W$, $\alpha \in D$, define $x_\gamma(\xi) = \phi(w)x_\alpha(\xi)\phi(w)^{-1}$.

The last assertions of (ii) follow from 11.1.4 and 11.2.7(ii). Finally, (iii) is easily checked.

11.3.7. <u>Corollary</u>. <u>The structure constants</u> $c_{\alpha,\beta;i,j}$ (<u>for some fixed order of</u> R) <u>are determined up to equivalence by the root system</u> R.

In the next chapter (12.12, see the remark after 12.14) we shall see how to obtain the structure constants from the root system in the simply laced case.

11.3.8. We terminate this section with a variant of 11.3.2. Notations are as in 11.3.1. Define $\tilde{T}$ as before. Assume we are given for each <u>positive</u> root γ and $\xi \in k$ a symbol $\tilde{x}'_\gamma(\xi)$, such that the analogues (1)', (2)' and (3)' of (1), (2), (3) hold. Assume, moreover that for $\alpha \in D$ we are given a symbol $\tilde{n}'_\alpha$, such that the analogues (5)' and (6)' of (5) and (6) hold, and such that

(7)' $\tilde{x}'_\alpha(\xi)\tilde{n}'_\alpha\tilde{x}'_\alpha(\xi^{-1})\tilde{n}'^{-1}_\alpha\tilde{x}'_\alpha(\xi) = \tilde{\alpha}^\vee(\xi)\tilde{n}'_\alpha \quad (\alpha \in D, \xi \in k^*)$.

Let $\tilde{G}'$ be the group generated by $\tilde{T}$, the $\tilde{x}'_\gamma(\xi)$ ($\gamma > 0, \xi \in k$) and $\tilde{n}'_\alpha$ ($\alpha \in D$), subject to the relations (1)'...(7)' and moreover

(8)' $\tilde{n}'_\alpha\tilde{t}\tilde{n}'^{-1}_\alpha = s_\alpha(\tilde{t}) \quad (\tilde{t} \in \tilde{T}, \alpha \in D)$.

There is an obvious homomorphism $\pi' : \tilde{G}' \to G$.

11.3.9. <u>Proposition</u>. π' <u>is an isomorphism of abstract groups</u>. Let $\tilde{X}'_\gamma$ ($\gamma > 0$) be the subgroup of $\tilde{G}$ generated by the $\tilde{x}'_\gamma(\xi)$ ($\xi \in k$). As in the proof of 11.3.2, one sees that $\tilde{U}' = \prod_{\gamma>0} \tilde{X}'_\gamma$ (the product taken in any order) is a subgroup of $\tilde{G}'$, mapped isomorphically onto the corresponding subgroup U of G by π'.

Let $\tilde{B}'$ be the subgroup of $\tilde{G}'$ generated by $\tilde{T}$ and $\tilde{U}'$, then π' maps $\tilde{B}'$ isomorphically onto B.

If $w \in W$, and $w = s_{\alpha_1} \dots s_{\alpha_h}$ is a shortest expression, put $\tilde{\phi}'(w) = \tilde{n}'_{\alpha_1} \dots \tilde{n}'_{\alpha_h}$, this is independent of the choice of the shortest expression. Put $\tilde{C}'(w) = \tilde{B}'\tilde{\phi}'(w)\tilde{B}'$. As before, one shows that $\tilde{G}' = \bigcup_{w \in W} \tilde{C}'(w)$, and 11.3.9 follows.

11.4. The isomorphism theorem.

The following elementary lemma will be needed.

11.4.1. Lemma. *Let G and G_1 be algebraic groups, and $\phi: G \to G_1$ a homomorphism of abstract groups. Assume there exists a non-empty open subset U of G such that the restriction of ϕ to U is a morphism of varieties. Then ϕ is a homomorphism of algebraic groups.*

If $g \in G$ the restriction of ϕ to the left translate g.U is also a morphism of varieties, since left translations are isomorphisms of varieties. Because the translates g.U cover G, the assertion follows.

11.4.2. Now assume that G and G_1 are two connected reductive linear algebraic groups, with maximal tori T, T_1 and corresponding root data $\Psi = (X, X^\vee, R, R^\vee)$, $\Psi_1 = (X_1, X_1^\vee, R_1, R_1^\vee)$, respectively. Assume that $\phi: G \to G_1$ is a surjective homomorphism of algebraic groups, with finite kernel. ϕ is called an *isogeny*. Then Ker ϕ is a central subgroup of G (see 2.2.2(3)), hence lies in T (9.3.5(iii)). It also follows that dim G = dim G_1. Also, ϕ defines a homomorphism of character groups $f: X_1 = X^*T_1 \to X = X^*T$ and a homomorphism $f^\vee: X^\vee \to X_1^\vee$ of groups of 1-parameter subgroups with $f^\vee u = f \circ u$. We have $\langle x_1, f^\vee u\rangle = \langle fx_1, u\rangle$ $(x_1 \in X_1, u \in X^\vee)$. Now let $\alpha \in R$. If X_α is as in 9.3.6 then ϕX_α is a connected unipotent 1-dimensional subgroup of G_1 which is normalized but not centralized by T_1. By 9.3.9(1) there is $\alpha_1 \in R_1$ with $\phi X_\alpha = X_{\alpha_1}$. Conversely, if $\alpha_1 \in R_1$ then $(\phi^{-1}X_{\alpha_1})^0$ is a connected unipotent 1-dimensional subgroup of G, normalized by T, hence one of the X_α. It follows that there is a bijection $u: R \xrightarrow{\sim} R_1$ with $\phi X_\alpha = X_{u\alpha}$.

Now let ϕ be an isomorphism of algebraic groups. It is then clear that $f(u\alpha) = \alpha$ $(\alpha \in R)$. Hence f defines an isomorphism of root data $\Psi_1 \to \Psi$, i.e. an isomorphism $X_1 \xrightarrow{\sim} X$ which maps R_1 onto R and is such that the dual isomorphism $X^\vee \to X_1^\vee$ maps $R^\vee$ onto $R_1^\vee$. We write $f = f(\phi)$.

11.4.3. <u>Theorem</u> (<u>Isomorphism theorem</u>). <u>Let</u> f <u>be an isomorphism of</u> Ψ_1 <u>onto</u> Ψ. <u>There exists an isomorphism of algebraic groups</u> $\phi: G \to G_1$ <u>such that</u> $\phi T = T_1$ <u>and</u> $f = f(\phi)$. <u>If</u> ϕ' <u>is another isomorphism with this property, there exists</u> $t \in T$ <u>such that</u> $\phi' x = \phi(txt^{-1})$ $(x \in G)$.

Let $(x_\alpha)_{\alpha \in R}$ and $(y_{\alpha_1})_{\alpha_1 \in R_1}$ be realizations of R, R_1, respectively. By 11.3.6 we may assume that the corresponding structure constants $c_{\alpha,\beta;i,j}$ and $c_{f^{-1}\alpha, f^{-1}\beta;i,j}$ are equal, whenever they are defined (assuming suitable orderings of R and R_1).

Denote by $\tilde{G}$ and $\tilde{G}_1$ groups defined as in 11.3.1, generated by $\tilde{T} = \mathrm{Hom}(X, k^*)$, $\tilde{x}_\gamma(\xi)$ $(\gamma \in R, \xi \in k)$ and $\tilde{T}_1 = \mathrm{Hom}(\tilde{X}_1, k^*)$,

$\tilde{y}_{\gamma_1}(\xi)$ $(\gamma_1 \in R_1, \xi \in k)$, respectively. It is then clear that there exists an isomorphism of abstract groups $\tilde{\phi}\colon \tilde{G} \to \tilde{G}_1$ with $(\tilde{\phi}\tilde{t})(x_1) = \tilde{t}(fx_1)$ $(\tilde{t} \in \tilde{T}, x_1 \in X_1)$ and $\tilde{\phi}\tilde{x}_\gamma(\xi) = \tilde{y}_{f^{-1}\gamma}(\xi)$ $(\gamma \in R, \xi \in k)$. Then $\tilde{\phi}\tilde{n}_\gamma = \tilde{n}_{f^{-1}\gamma}$.

Let π and π_1 be the isomorphisms $\tilde{G} \to G$, $\tilde{G}_1 \to G_1$ of 11.3.2. Then $\phi = \pi_1 \circ \tilde{\phi} \circ \pi^{-1}$ is an isomorphism of G onto G_1. But it follows from the definition of $\tilde{\phi}$ that the restriction of ϕ to the big all $C(w_0)$ relative to a Borel subgroup $B \supset T$ (see 10.2.10) is a morphism of varieties (check this). By 11.4.1 it follows that ϕ is a homomorphism of algebraic groups. Reversing the roles of G and G_1 we conclude that ϕ is an isomorphism of algebraic groups. That $f = f(\phi)$ follows from the definitions.

If ϕ' is a second isomorphism with $f = f(\phi')$, then $\psi = \phi^{-1} \circ \phi'$ is an isomorphism of algebraic groups $G \to G$, with $\psi t = t$ $(t \in T)$ and $\psi X_\gamma = X_\gamma$, for all $\gamma \in R$. So $\psi(x_\gamma(\xi)) = x_\gamma(c_\gamma \xi)$, with $c_\gamma \in k^*$ and $c_\gamma c_{-\gamma} = 1$ (by 11.2.1). Let D be a basis of R, for some positive system of roots R^+. We can then find $t \in T$ such that $\alpha(t) = c_\alpha$ $(\alpha \in D)$. It follows that $\psi x = txt^{-1}$. if either $x \in T$, or $x = x_\gamma(\xi)$ with $\pm\gamma \in D$. We conclude from 9.3.6(ii) that $\psi x = txt^{-1}$ for all $x \in G$. This establishes the last point of the theorem.

11.4.4. For later use we discuss an application of the previous theorem. Fix a system of positive roots R^+, let D be the corresponding basis and $\mathcal{D}$ the Dynkin diagram. Let $D = D_1 \cup D_2$ be a decomposition of D with the properties of 11.3.5.

We make the following assumptions:

(a) G is semi-simple and of adjoint type (see 9.4.2). This means that $X = Q$, the lattice spanned by R.

(b) $\mathcal{D}$ is simply laced (no multiple bonds occur).

(c) There exists an automorphism f of $\mathcal{D}$ such that $fD_1 = D_1$, $fD_2 = D_2$.

Since D is a basis of Q (because of (a)), f extends to an automorphism of Q, also denoted by f. It is clear that f defines an isomorphism of the root datum $\Psi(G) = (Q,Q^{\vee},R,R^{\vee})$ onto itself. Let h be the order of f.

11.4.5. Corollary. <u>There exists an automorphism ϕ of the algebraic group G such that $f(\phi) = f$ and that ϕ has order h.</u>

The first point follows from 11.4.3. It also follows that we then have $\phi^h x = txt^{-1}$, for some $t \in T$. We shall now show that, under our assumptions, we can achieve that $t = 1$.

The assumptions show that, if the realization (x_γ) of R is such that the structure constants satisfy 11.3.6(a), we have $c_{\alpha,\beta;1,1} = c_{f\alpha,f\beta;1,1}$, if $\alpha,\beta \in D$ are non-orthogonal. It follows from the proof of 11.4.3 (check this) that then we may assume ϕ to be such that $\phi x_\alpha(\xi) = x_{f^{-1}\alpha}(\xi)$ $(\alpha \in D, \xi \in k)$. Hence $\phi^h x_\alpha(\xi) = x_\alpha(\xi)$ $(\alpha \in D)$. If t is as above, then $\alpha(t)=1$ for all $\alpha \in D$, whence $t = 1$ (since D spans Q).

11.4.6. So far, we did not need the classification of the Dynkin diagrams to irreducible root systems. At this point it is useful to recapitulate this classification, to illustrate the last result. We refer to [8, Ch.VI, p.250-275] for details about the classification.

The Dynkin diagrams of the irreducible, simply laced root sytems are:

A_ℓ ($\ell \geq 1$)

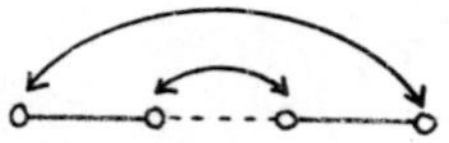

D_ℓ ($\ell \geq 4$)

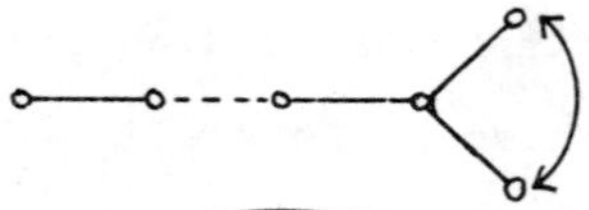

E_6

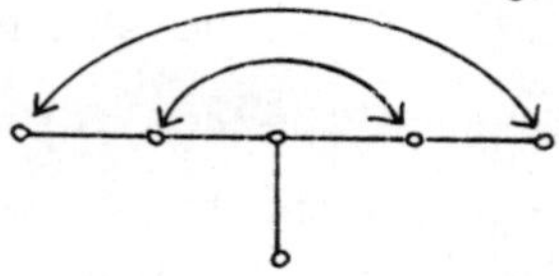

E_7

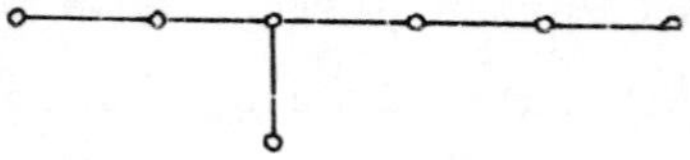

E_8

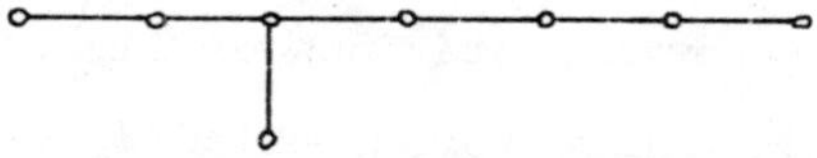

The arrows indicate the automorphisms f of order 2. Besides these, there exists in type D_4 automorphisms of order 3, as is clear from the diagram.
These exhaust all the possible automorphisms. They satisfy the condition (c) of 11.4.4, except for the case A_ℓ with ℓ even.

The Dynkin diagrams with multiple bonds are:

B_ℓ ($\ell \geq 2$)

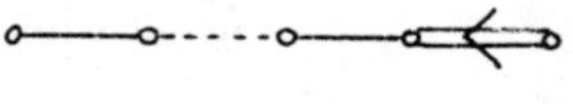

C_ℓ ($\ell \geq 3$)

F_4 

G_2 

These can be obtained from simply laced diagrams by "folding", using an automorphism f. We shall deal with this in the next chapter (see 12.17). There we shall use folding and 11.4.5, to reduce the existence theorem (of root data with a given root system) to the simply laced case.

11.4.7. There is a generalization of the isomorphism theorem, dealing with isogenies. We shall discuss it briefly.
Let $\phi: G \to G_1$ be an isogeny of connected, reductive linear algebraic groups. We use the notations of 11.4.2. First notice that f is an injective homomorphism $X_1 \to X$ and that fX_1 has finite index in X (the index being equal to the order of Ker ϕ). Denote by $(x_\alpha)_{\alpha \in R}$ and $(y_{\alpha_1})_{\alpha_1 \in R_1}$ realizations in G and G_1, respectively. Then $\phi x_\alpha(\xi) = y_{u\alpha}(h(\xi))$, where $h \in k[T]$ and it follows that $h(\xi+\eta) = h(\xi) + h(\eta)$. If the characteristic p of k is 0, then $h(\xi) = a\xi$, $a \in k^*$ and if $p > 0$ then h is a p-polynomial (see 2.6.3). Moreover, from $tx_\alpha(\xi)t^{-1} = x_\alpha(\alpha(t)\xi)$, we conclude that $h(\alpha(t)\xi) = h((u\alpha)(\phi(t)\xi)$, if $t \in T, \xi \in k$. It follows that there exist (if $p > 0$) p-powers $q(\alpha)$ such that $f(u\alpha) = q(\alpha)\alpha$. It also follows that $f^\vee(\alpha^\vee) = q(\alpha)(u\alpha)^\vee$ (use 11.2.1). If $p = 0$ we define $q(\alpha) = 1$ for all $\alpha \in R$. The preceding formulas still hold in that case.

We say that f,u and the function $q: R \to \{p^n\}_{n>0}$ (p denoting the characteristic exponent of k) define a p-morphism of Ψ_1 to Ψ if the properties stated above hold: f is an injective homomorphism $X_1 \to X$ with finite cokernel, u is a bijection $R \to R_1$ and $f(u\alpha) = q(\alpha)\alpha$, $f^\vee(\alpha^\vee) = q(\alpha)(u\alpha)^\vee$ $(\alpha \in R)$. If (f,u,q) comes, in the manner described above, from an isogeny ϕ, we write $f = f(\phi)$, $u = u(\phi)$, $q = q(\phi)$.

As an example of a p-morphism when $p > 0$ we mention the Frobenius morphisms: $\Psi = \Psi_1$ and $f = q.\mathrm{id}$, where q is a p-power.

The next lemma gives some properties of p-morphisms.

11.4.8. Lemma. (i) If $\alpha,\beta \in R$ then $q(\beta)\langle u\alpha,(u\beta)^\vee\rangle = q(\alpha)\langle\alpha,\beta^\vee\rangle$;

(ii) We have $u(s_\alpha\beta) = s_{u\alpha}(u\beta)$ $(\alpha,\beta \in R)$;

(iii) If $\alpha \in R$, $w \in W$ then $q(w\alpha) = q(\alpha)$;

(iv) If $\alpha,\beta \in R$, $\langle\alpha,\beta^\vee\rangle > 0$ and $q(\alpha) < q(\beta)$ then $p = 2$ or 3 and $\langle\alpha,\beta^\vee\rangle = p$, $\langle u\alpha,(u\beta)^\vee\rangle = 1$.

We have

$$q(\beta)\langle u\alpha,(u\beta)^\vee\rangle = \langle u\alpha,f^\vee(\beta^\vee)\rangle = \langle f(u\alpha),\beta^\vee\rangle = q(\alpha)\langle\alpha,\beta^\vee\rangle,$$

proving (i). Also

$$f(s_{u\alpha}(u\beta)) = f(u\alpha-\langle u\alpha,(u\beta)^\vee\rangle u\beta) =$$

$$q(\alpha)\alpha-q(\beta)\langle u\alpha,(u\beta)^\vee\beta\rangle =$$

$$q(\alpha)(s_\alpha\beta)(\text{by (i)}) = q(\alpha)q(s_\alpha\beta)^{-1}f(u(s_\alpha\beta)).$$

It follows from the injectivity of f that the roots $s_{u\alpha}(u\beta)$

and $u(s_\alpha\beta)$ of R_1 differ by the positive factor $q(\alpha)q(s_\alpha\beta)^{-1}$. Hence these roots must be equal, and $q(s_\alpha\beta) = q(\alpha)$. This establishes (ii), and (iii) also follows. Finally, (iv) follows from (i), using 9.2.1.

11.4.9. Theorem (Isogeny theorem). *Let (f,u,ϕ) be a p-morphism of Ψ_1 to Ψ. There exists an isogeny $\phi: G \to G_1$ with $\phi T = T_1$ such that $f = f(\phi)$, $u = u(\phi)$, $q = q(\phi)$. If ϕ' is another isogeny with these properties, there exists $t \in T$ such that $\phi' x = \phi(txt^{-1})$ $(x \in G)$.*

Let $(x_\alpha)_{\alpha\in R}$ and $(y_{\alpha_1})_{\alpha_1\in R_1}$ be realizations of R and R_1, respectively. We may assume that the corresponding structure constants have the properties of 11.3.6(ii).

Denote by $\tilde{G}$ and $\tilde{G}_1$ groups defined as in 11.3.1. If we establish that there is a homomorphism $\tilde{\phi}: \tilde{G} \to \tilde{G}_1$ such that $(\tilde{\phi}\tilde{t})(x_1) = \tilde{t}(fx_1)$ $(\tilde{t} \in \tilde{T}, x_1 \in X_1)$, $\tilde{\phi}\tilde{x}\,(\xi) = \tilde{y}_{u\gamma}(\varepsilon_\gamma \xi^{q(\gamma)})$, with suitable $\varepsilon_\gamma = \pm 1$ $(\gamma \in R, \xi \in k)$ then we can proceed as in the proof of 11.4.3 to prove the existence of ϕ.

In the particular case that $G = G_1$ and the p-morphism is a Frobenius morphism the existence of a homomorphism $\tilde{\phi}$ with $\tilde{\phi}\tilde{x}_\gamma(\xi) = \tilde{x}_\gamma(\xi^q)$ follows from the fact that the structure constants lie in the prime field.

In the general case, the essential part of the proof of the existence of $\tilde{\phi}$ is the verification that the relations (2) of 11.3.1 are preserved. But this verification can be carried out in rank 2 root systems. This reduces things to the case that R has rank 2. We may also assume G to be semi-simple, of adjoint type (since we are dealing with properties of the

$\tilde{x}_\gamma$). Except for the case that the situation of 11.4.8(iv) arises, we only have to deal with Frobenius morphisms. This means that we are reduced to considering the case that $G = G_1$ is semi-simple, adjoint and that R has type B_2 $(p = 2)$ or G_2 $(p = 3)$. In these cases one can use 11.3.9. We shall only say a few words about the last case, which is the most complicated one. Let $\tilde{G}'$ be the group defined in 11.3.8. We now have $u\alpha = \beta$, $u\beta = \alpha$, $u(\alpha+\beta) = 3\alpha+\beta$, $u(2\alpha+\beta) = 3\alpha+2\beta$, $u(3\alpha+\beta) = \alpha+\beta$, $u(3\alpha+2\beta) = 2\alpha+\beta$. Moreover $q(\alpha)=3^{n+1}$, $q(\beta)=3^n$. It suffices to deal with the case $n = 0$ (the general case is obtained by composing with a Frobenius morphism). Then $q(\gamma)=3$ if γ is short and $q(\gamma)=1$ if γ is long. We have to show that there exists a homomorphism $\tilde{\phi}': \tilde{G}' \to \tilde{G}'$ with $\tilde{\phi}'\tilde{x}'_\gamma(\xi) = \tilde{x}'_{u\gamma}(\varepsilon_\gamma \xi^{q\gamma})(\gamma > 0)$, $\tilde{\phi}'\tilde{n}'_\alpha = \beta^\vee(\varepsilon_\alpha)\tilde{n}'_\beta$, $\tilde{\phi}'\tilde{n}'_\beta = \alpha^\vee(\varepsilon_\beta)\tilde{n}'_\alpha$, the $\varepsilon_\gamma = \pm 1$ being suitable signs. Now the analysis of the G_2-case made in the proof of 11.2.7 shows that the group $\tilde{U}'$ generated by the subgroups $\tilde{X}'_\gamma$ is isomorphic to $k^2 \times M$, where $M = k^4$, the multiplication being given by

$$(\xi,\eta,m)(\xi',\eta',m') = (\xi+\xi',\eta+g(-\xi,w') + f(\rho\begin{pmatrix}1 & -\xi'\\ 0 & 1\end{pmatrix}m,m'),\rho\begin{pmatrix}1 & -\xi'\\ 0 & 1\end{pmatrix}m+m'),$$

where $\xi,\eta,\xi',\eta' \in k$, $m,m' \in M$, and ρ,g,f are as in the proof of 11.2.7. The desired homomorphism can be described explicitly by a map ψ of $k^2 \times M$. A computation will show that the signs can be chosen so that ψ is a homomorphism. The existence of ϕ then follows as before. The proof of the uniqueness statement is similar to the proof of the corresponding part of 11.4.3.

11.4.10. Exercise. Complete the details of the proof of 11.4.9.

For more details about "exceptional" isogenies we refer to [34, §11].

Notes.

Proofs of the main result of this chapter, the isomorphism theorem 11.4.2, can also be found in [11 , exp.24], [14 , exp. XXIII], [23 , no.33]. In [11], [23] this is done for semi-simple groups, and in [31 , no.2] it is shown how one can reduce the case of a general reductive group to the semi-simple case. In [14] a more general result is established, over a general base scheme.

We have not assumed here knowledge about structure constants of semi-simple Lie algebras.

11.1. contains familiar material about root systems of rank 2.

In 11.2 structure constants are introduced, and they are discussed in detail in the case of semi-simple rank 2. Our treatment is inspired by the computations made by Demazure in [14 , exp.XXIII, no.3] (reproduced in [23 , p.209-215]). However, there the computations are used to normalize structure constants, whereas here they are used to deduce properties of and relation between structure constants. We have tried to bring out clearly the fact, an analogue of which is familiar in the theory of semi-simple Lie algebras, that properties of low-dimensional representations of $\mathbb{SL}_2$ are responsible for relations like those of 11.2.5(i) (see the proof of 11.2.5). Nevertheless the most complicated case G_2 still requires some unpleasant computations.

11.2.8 is implicit in the results of [14 , exp.XXIII, no.3], but is not explicitly stated there.

To pass from rank 2 to arbitrary rank we use in 11.3 a weak version of results of Steinberg on generators and relations for semi-simple groups, generalized to the reductive case. For Steinberg's results see [32] or [34].

The version we give is theorem 11.3.2. Implicit in it is a result of Curtis which states, roughly speaking, that our group $\widetilde{G}$ can be obtained by "amalgamation" of subgroups corresponding to the root systems $R_{\alpha,\beta}$ $(\alpha,\beta \in D, \alpha \neq \beta)$. See [12].

The fact that the structure constants are, up to equivalence, completely determined by the root system (see 11.3.7) has not been stated explicitly in the literature, as far as I know. A different proof has been given by Hassan Azad (see [1].

It should be noted that the proof of the isomorphism theorem 11.4.2 involves only very little algebraic geometry (only the easy lemma 11.4.1). The difficulties of the proof come in the analysis of the group-theoretical structure of reductive groups.

The isogeny theorem is due to Chevalley, it is also proved in [11] and [14].

12. The existence theorem.

This last chapter is devoted to the proof of the existence of a reductive group with given root datum. The existence theorem is as follows.

12.1. *Theorem* (*Existence theorem*). *Let* $\Psi = (X,R,X^{\vee},R^{\vee})$ *be a root datum. There exists a connected reductive linear algebraic group* G, *containing a maximal torus* T *such that the root datum* $\Psi(G,T)$ *is isomorphic to* Ψ.

We know from 11.4.2 that such a G is unique, up to an isomorphism of algebraic groups.

12.2. We first make some preliminary comments on the proof of the existence theorem. From the results of 11.3 we see that, if the root datum Ψ is given, one can define a group $\tilde{G}$ by generators and relations which is isomorphic to a group G with the property of 12.1. However, this presupposes some information about structure constants. Moreover, if one has such information, one must establish that the abstract group $\tilde{G}$ is non-trivial. To deal with this point one has to use a representation of $\tilde{G}$.

We shall proceed in the following manner:

(a) The existence proof for G will be reduced in 12.7 to the construction of a linear group satisfying certain conditions. These are suggested by 11.3.9.

(b) The reduction theorem 12.8 reduces the proof of 12.1 to the proof of the existence of a semi-simple adjoint group with a given root system R.

(c) The latter existence result will first be dealt with in the case that R is simply laced. G will be constructed (using 12.7) as a group of linear transformations of its Lie algebra $\mathfrak{g}$. To do this we first have to construct this Lie algebra. The information about structure constants which is required for this construction can nowadays be obtained quite easily (see 12.14).

(d) The case of a general root system R is reduced to the simply laced case by using automorphisms (see 12.18).

We shall now carry out the details of this programme.

12.3. Assume that $\Psi = (X,R,X^{\vee},R^{\vee})$ is a root datum. First observe that if $R = \phi$, the existence theorem is easy. For then G must be a torus whose character group is X, whose existence follows from 2.5.7.

From now on assume $R \neq \phi$. Let W be the Weyl group of Ψ, i.e. the group generated by the reflections s_{α} ($\alpha \in R$), it is a finite group (9.1.10(1c)). Put $X_{\mathbb{R}} = X \otimes_{\mathbb{Z}} \mathbb{R}$ and fix a positive definite symmetric bilinear form on $X_{\mathbb{R}}$ which is W-invariant (9.1.8). Also put $V = Q \otimes_{\mathbb{Z}} \mathbb{R}$, as in 9.1.6. Then R is a root system in V.

We shall use the properties of systems of positive roots etc. established in the preceding chapters. However, there some of these properties were not proved for abstract root systems, but only under the assumption that R was the root system of a reductive group. This is not a serious problem, as we shall now show.

Put $V' = \{x \in V \mid (x,\alpha) \neq 0 \text{ for all } \alpha \in R\}$. For convenience we now define a <u>system of positive roots</u> $R^+ \subset R$ to be a set $\{\alpha \in R \mid (x,\alpha) > 0\}$, for some $x \in V'$. Then R^+ has the properties enunciated in 9.2.2, as we observed there. Two systems of positive roots R^+ and $\tilde{R}^+$ are <u>adjacent</u> if the intersection $R^+ \cap \tilde{R}^+$ has one element less than R^+ and $\tilde{R}^+$.

12.4. <u>Lemma</u>. (i) <u>Let</u> R^+ <u>and</u> $\tilde{R}^+$ <u>be two systems of positive roots. There is a chain</u> $R^+ = R_0^+, R_1^+, \ldots, R_h^+ = \tilde{R}^+$ <u>of systems of positive roots such that</u> R_i^+ <u>and</u> R_{i+1}^+ <u>are adjacent</u> $(0 \leq i \leq h-1)$;

(ii) <u>If</u> R^+ <u>and</u> $\tilde{R}^+$ <u>are adjacent there is</u> $\alpha \in R^+$ <u>such that</u> $\tilde{R}^+ = s_\alpha R^+$.

The proof of (i) is an adaptation of that of 10.1.7 and can be omitted. To prove (ii), assume that R^+ and $\tilde{R}^+$ are defined by elements x and y of V, respectively, and let $R^+ \cap \tilde{R}^+ = R^+ - \{\alpha\}$. We then have $(y,\alpha) < 0$. Hence if $\gamma \in R^+$, $\gamma \neq \alpha$,

$$(\gamma, s_\alpha y) = (y, s_\alpha \gamma) = (y, \gamma - 2(\alpha,\alpha)^{-1}(\gamma,\alpha)\alpha) > 0,$$

whereas $(\alpha, s_\alpha y) = -(y,\alpha) > 0$.

It follows that $s_\alpha \tilde{R}^+ \subset R^+$, whence $\tilde{R}_+ = s_\alpha R^+$.

12.5. Fix a system of positive roots R^+ and define the basis D of R^+ to be the set of $\alpha \in R^+$ such that R^+ and $s_\alpha R^+$ are adjacent, as before. We then have again the properties of 10.1.9.

Denote by W' the subgroup of W generated by the reflections s_α with $\alpha \in D$. The proof of 10.1.10 shows that $R = W'.D$ and that 10.1.10(iii) holds. But it then also follows that $W' = W$: Take $\gamma \in R$ and write $\gamma = w'\alpha$ with $w' \in W$, $\alpha \in D$. Then $s_\gamma = w's_\alpha(w')^{-1} \in W'$, whence $W \subset W'$ and $W' = W$. We have thus obtained the results of 10.1.10. The results of 10.2.2 now also hold. Notice that these also imply the following: if $w \in W$ and $wR^+ = R^+$ then $w = 1$ (this we used implicitly in the preceding chapters, as a consequence of 7.3.7). We have now recovered, for abstract root systems, the results of the previous chapters.

12.6. Let R^+ and D be as before. Assume that A is a vector space over k and that we are given

(a) a torus $T \subset GL(A)$ with character group X,

(b) for each $\gamma \in R^+$ an isomorphism x_γ of $\mathbb{G}_a$ onto a closed subgroup X_γ of $GL(A)$,

(c) for each $\alpha \in D$ an element $n_\alpha \in GL(A)$, such that the following relations hold.

(1) There exist $c_{\gamma,\delta;i,j} \in k$ such that

$$(x_\gamma(\xi), x_\delta(\eta)) = \prod_{\substack{i\gamma+j\delta\in R^+ \\ i,j > 0}} x_{i\gamma+j\delta}(c_{\gamma,\delta;i,j}\xi^i\eta^j),$$

if $\gamma,\delta \in R^+$, $\xi,\eta \in k$ (the order in the product being given in advance),

(2) $tx_\gamma(\xi)t^{-1} = x_\gamma(\gamma(t)\xi)$ $(\gamma \in R^+, t \in T, \xi \in k)$,

(3) $n_\alpha^2 = t_\alpha \in T$ and $x(t_\alpha) = (-1)^{\langle x,\alpha^\vee\rangle}$ $(\alpha \in D, x \in X)$.

Let the Weyl group W act on T such that $x(w.t) = (w^{-1}x)(t)$ $(t \in T, x \in X, w \in W)$.

(4) $n_\alpha t n_\alpha^{-1} = s_\alpha . t$ $(\alpha \in D, t \in T)$,

(5) $x_\alpha(\xi) n_\alpha x_\alpha(\xi^{-1}) n_\alpha^{-1} x_\alpha(\xi) = \alpha^\vee(\xi) n_\alpha$ $(\alpha \in D, \xi \in k^*)$,

where $\alpha^\vee(\xi)$ is the element of T defined by $x(\alpha^\vee(\xi)) = \xi^{\langle x,\alpha^\vee \rangle}$, $\xi \in k^*, x \in X$.

The reader will recognize some of the relations of 11.3.8.
Denote by G the subgroup of GL(A) generated by T, the X_γ $(\gamma \in R^+)$, and the n_α $(\alpha \in D)$.

12.7. Proposition. *G is a connected reductive linear algebraic group whose root datum is* $\Psi = (X, R, X^\vee, R^\vee)$.

From (5) we see that G is also generated by the connected, closed subgroups T, X_γ $(\gamma \in R^+)$ and $n_\alpha X_\alpha n_\alpha^{-1}$ $(\alpha \in D)$ of GL(A). It follows from 6.1 that G is a connected, closed subgroup of GL(A).

If $\gamma \in R^+$, $\gamma = \sum_{\alpha \in D} n_\alpha \alpha$, the height $\mathrm{ht}\gamma$ of γ is Σn_α (see the proof of 11.3.2). Denote by U_n the (connected, closed) subgroup of G generated by the X_γ with $\gamma \in R^+$, $\mathrm{ht}\gamma \geq n$. Put $U = U_1$, $B = T.U$. By recurrence, using (1), one shows that the U_n form a descending set of closed normal subgroups of B, such that U_n/U_{n+1} is commutative and that, set-theoretically, $U_n = \prod_{\mathrm{ht}\gamma \geq n} X_\gamma$, for any order. In particular we have $U = \prod_{\gamma \in R^+} X_\gamma$. It follows that $\dim U \leq |R^+|$. But since the Lie algebra of U contains the nonzero vectors $dx_\gamma(1)$ $(\gamma \in R^+)$, which are weight vectors for T with weight γ, we are in the situation of 10.1.2. Then 10.1.2 shows that the product map $(\xi_\gamma)_{\gamma \in R^+} \mapsto \prod_{\gamma \in R^+} x_\gamma(\xi_\gamma)$ defines an isomorphism of $G_a^{R^+}$ onto U. In particular, $\dim U = |R^+|$.

Let $w \in W$ have a shortest expression $w = s_{\alpha_1}\dots s_{\alpha_h}$ ($\alpha_i \in D$). Put $\dot{w} = n_{\alpha_1}\dots n_{\alpha_h}$. Then $\dot{w}T$ and $C(w) = B\dot{w}B$ are independent of the choice of the shortest expression. Moreover $C(w)$ is a locally closed subset of G (being an orbit of the algebraic group $B \times B$), whose dimension is at most $|R|+r$ (where $r = \text{rank } X$), the maximal value being attained only for the longest element w_0. One concludes that $\dim G = |R|+r$.
Take $\gamma \in R$ and put $\gamma = w\alpha$ ($w \in W, \alpha \in D$). Then, $\dot{w}$ being as before, $x(\xi) = \dot{w}x_\alpha(\xi)\dot{w}^{-1}$ defines an isomorphism of G_a onto a closed subgroup of G, and $dx(1)$ is a nonzero weight vector of T, for the weight γ. Hence $\mathfrak{g}$ contains weight vectors for all $\gamma \in R$. Since $\dim \mathfrak{g} = |R|+r$, it follows that the corresponding weight spaces $\mathfrak{g}_\gamma$ must be 1-dimensional, and are uniquely determined (i.e. independent of the choice of w and α).
Now take $\alpha \in D$ and put $x_{-\alpha}(\xi) = n_\alpha x_\alpha(\xi)n_\alpha^{-1}$, then $\text{Im } dx_{-\alpha} = \mathfrak{g}_{-\alpha}$. It is clear that $H = Z_G(\text{Ker } \alpha)^0$ contains T, X_α and $X_{-\alpha}$. Using 4.4.7, we conclude from the information about $\mathfrak{g}$ which we just found that $\mathfrak{h}$ contains $\mathfrak{t}, \mathfrak{g}_\alpha$ and $\mathfrak{g}_{-\alpha}$. Because H is connected (7.3.5(i)) it follows that H is generated by T, X_α and $X_{-\alpha}$. The relations (2) and (5) imply that H cannot be solvable (check this). Consequently, α is a root of G with respect to T (see the definition of roots in 9.1.4). But then all $\gamma \in R$ are roots. Since $\mathfrak{g} = \mathfrak{t} \oplus \sum_{\gamma \in R} \mathfrak{g}_\gamma$, the unipotent radical of G must be trivial. So G is reductive.
It follows from (5) that $\xi \mapsto \alpha^\vee(\xi)$ defines the coroot corresponding to the root α. One then readily concludes that Ψ

is indeed the root datum of (G,T). This proves 12.7.

The next result will enable us to reduce the proof of the existence theorem to the case of semi-simple adjoint groups. Let $\Psi = (X,R,X^\vee,R^\vee)$ be a root datum. Assume that X_1 is a sublattice of X of finite index containing R. Then we can view $X^\vee$ as a sublattice of finite index of the dual $X_1^\vee$, so $R^\vee$ is a subset of $X_1^\vee$. Thus $\Psi_1 = (X_1,R,X_1^\vee,R^\vee)$ is a root datum. We prefer to write $\Psi_1 = (X_1,R_1,X_1^\vee,R_1^\vee)$.

12.8. Proposition. *Assume that there exists a connected reductive group* G_1, *together with a maximal torus* T_1, *such that* $\Psi(G_1,T_1) = \Psi_1$. *Then there exists a similar pair* (G,T) *with* $\Psi(G,T) = \Psi$.

We may assume that the index of X_1 in X is a prime number ℓ.

Let $(x'_\gamma)_{\gamma \in R}$ be a realization of R in G_1. Fix a system of positive roots R^+ in R, let D be the corresponding basis and $B_1 = T_1U$ the corresponding Borel subgroup of G_1. Denote by $\Omega_1 = T_1Y$ the big cell in G_1 defined by B_1. Here Y = UaU, with a suitable element $a \in G_1$.

We can construct a torus T with character group X. The injection $X_1 \to X$ defines a surjective homomorphism of tori $\pi: T \to T_1$ (use 2.5.7). The Weyl group W operates on T.

If G exists, the big cell in G (defined by R^+) must be isomorphic to T × Y. It follows that the quotient field k(G) of k[G] must be isomorphic to the quotient field F of k[T × Y]. Since k(T) is an algebraic extension of degree ℓ

of $k(T_1)$ (because X_1 has index ℓ in X) it follows that we can view the field $k(G_1)$ as a subfield of F. Then F has degree ℓ over $k(G_1)$.

We shall now define G as a group of k-automorphisms of F. For $\gamma \in R, \xi \in k$ define an automorphism $x_\gamma(\xi)$ of $T \times Y$ by

$$x_\gamma(\xi)(t,y) = (t, x'_\gamma(\gamma(t)^{-1}\xi)y).$$

Also, T acts on $T \times Y$ by translations on the first factor. These automorphisms of $T \times Y$ define k-automorphisms $x_\gamma(\xi)$ and t of F.

Next let $\alpha \in D$ be a simple root. Put $n'_\alpha = x'_\alpha(1)x'_{-\alpha}(-1)x'_\alpha(1)$ and let Y_α, $\xi_\alpha y$, ϕ_α be as in 11.2.3. Define an automorphism n_α of $T \times Y_\alpha$ by

$$n_\alpha(t,y) = ((s_\alpha t)\alpha^\vee(-\xi_\alpha y)^{-1}, \phi_\alpha y)$$

(compare with 11.2.3). Then n_α defines a k-automorphism of F, also denoted by n_α. It is clear that the automorphisms of F just defined stabilized the subfield $k(G_1)$ of F and that, for $f \in k(G_1)$,

$$x_\gamma(\xi)f = x'_\gamma(\xi)f,\ tf = \pi(t)f,\ n_\alpha f = n'_\alpha f,$$

the actions in the right-hand sides coming from left translations by elements of G_1.

Denote by G the group of automorphisms of F generated by the automorphisms $x_\gamma(\xi)$ $(\gamma \in R, \xi \in k)$, $t \in T$, n_α $(\alpha \in D)$. We check that the relations (1),...(5) of 12.6 hold. This is trivial for all but the last one. To prove this relation (5), it suffices to check that

$$x_\alpha(\xi)n_\alpha x_\alpha(\xi^{-1})n_\alpha^{-1}x_\alpha(\xi)(1,y) = \alpha^\vee(\xi)n_\alpha(1,y),$$

for $y \in Y_\alpha$.

That the second components of both sides equal follows because these components can be computed in G_1. The equality of the first components follows by a straightforward computation, using 11.2.3 (check this).

We shall now produce a finite dimensional k-subspace A of F which is G-stable and satisfies the conditions of 12.6. First choose a finite dimensional G_1-stable subspace A' of $k[G_1]$ which generates the algebra $k[G_1]$ and is such that the representation of G_1 on A' defines an isomorphism of G_1 onto its image.

Next let $x \in X-X_1$, then $\ell x \in X_1$. Denote by f the corresponding element of $k[T] \subset F$. Then $f^\ell \in k[T_1] \subset k(G_1)$, and $F = k(G_1)(f)$. Choose $a \in k[G_1]$ such that $(af)^\ell \in k[G_1]$. By 2.3.4 all elements $(g.(af))^\ell$ $(g \in G)$ lie in a finite dimensional subspace of $k[G_1]$. Then the same is true for the elements g(af). In fact, we have $af \in k[T \times Y]$, and the latter algebra is isomorphic to an algebra $R = k[T_1,\dots,T_m,U_1,\dots,U_n,U_1^{-1},\dots,U_n^{-1}]$. It is easily seen, by looking at degrees, that the set of $r \in R$ such that r^ℓ lies in a given finite dimensional subspace of R spans a finite dimensional subspace.

Let $A'' \subset F$ be a finite dimensional G-stable subspace of F containing af and put $A = A' + A''$. Since A generates the field F over k, the action of G on F is completely determined by the representations of G in A. We identify G with its image in GL(A) and we shall show that the conditions of 12.6 are verified. The only things left are the proofs that $\rho(T) \subset GL(A)$ is indeed a torus with character groups X and that $\xi \mapsto \rho(x_\gamma(\xi))$ in GL(A) defines an isomorphism of $\mathbb{G}_a$ onto a closed subgroup of GL(A). The last point follows readily from the definitions. To prove the other point, observe that, by the definition of A', there is a set of nonzero weight vectors of T in A' whose corresponding weights span X_1. Also, by the definition of f, there is a nonzero weight vector for T in A" whose weight lies in $X-X_1$. Since X_1 has prime index in X, it follows that the weights of T in A span X, which is what we wanted. So the conditions of 12.6 are satisfied and 12.8 follows from 12.7.

12.9. Exercise. (a) Let G_1 be connected and semi-simple. Apply 12.8 to show that there exists a connected, semi-simple and simply connected group G with the same root system as G_1.
(b) Take $G_1 = \mathbf{SO}_n$ ($n \geq 5$, characteristic not 2). Show that G_1 is not simply connected. The corresponding group G of (a) is the spin group Spin (n).

12.10. Simply laced root systems.

Let R be a root system in the real vector space V. Denote the Weyl group by W. Recall that R is said to be simply _laced_ if $\langle\alpha,\beta^\vee\rangle = 0,\pm 1$ for any two linearly independent roots $\alpha,\beta \in R$ (11.3.4). This also means that all root strings have length ≤ 1 (see 11.1.6). We denote by (,) a positive definite symmetric bilinear form on V which is W-invariant, then $\langle\alpha,\beta^\vee\rangle = 2(\beta,\beta)^{-1}(\alpha,\beta)$.

12.11. _Lemma_. _Assume_ R _simply laced_, _let_ $\alpha,\beta \in R$.
(i) _If_ $\alpha \neq \pm\beta$ _and_ $\varepsilon = \pm 1$ _then_ $\alpha+\varepsilon\beta \in R$ _if and only if_ $\langle\alpha,\beta^\vee\rangle = -\varepsilon$;
(ii) _If_ R _is irreducible there is_ $w \in W$ _with_ $w\alpha = \beta$;
(iii) _If_ $\alpha,\beta,\alpha+\beta \in R$ _then_ $(\alpha+\beta)^\vee = \alpha^\vee + \beta^\vee$.
(i) readily follows from the proof of 9.2.3. The assertion of (ii) also follows from that proof if $(\alpha,\beta) \neq 0$. If $(\alpha,\beta) = 0$, the irreducibility of R (see 11.3.4) implies that there is a chain $\alpha = \alpha_0,\alpha_1,\ldots,\alpha_h = \beta$ of roots such that $(\alpha_i,\alpha_{i+1}) \neq 0$ (check this). The assertion (ii) then follows.
If $\alpha,\beta,\alpha+\beta \in R$ then $(\alpha,\beta) \neq 0$ by (i) and we know from the proof of (ii) that β and $\alpha+\beta$ are in the W-orbit of α. Hence $(\alpha,\alpha) = (\beta,\beta) = (\alpha+\beta,\alpha+\beta)$. Since we may identify $\alpha^\vee$ with $2(\alpha,\alpha)^{-1}\alpha$, we obtain (iii).

12.12. Assume R to be simply laced. It follows from 12.11(ii) that we may assume (,) to be such that $(\alpha,\alpha) = 2$ for all roots α. Then $(\alpha,\beta) = 0,\pm 1$ for any pair of linearly independent roots α,β.
Denote $Q \subset V$ the lattice spanned by R. Any basis D of R is a basis of Q, from which it follows that (x,x) is an even

integer for all $x \in Q$. Denote by f a bilinear $\mathbb{Z}$-valued form on Q such that

$$\begin{cases} (x,y) \equiv f(x,y) + f(y,x) \pmod 2 \\ \frac{1}{2}(x,x) \equiv f(x,x) \pmod 2, \end{cases}$$

for all $x,y \in Q$. Such f exist: take any basis $(e_i)_{1 \leqslant i \leqslant n}$ of Q and define f by $f(e_i,e_j) = (e_i,e_j)$ if $1 \leqslant i < j \leqslant n$, $f(e_i,e_i) = \frac{1}{2}(e_i,e_i)$, $f(e_i,e_j) = 0$ if $i > j$.

Fix a system of positive roots R^+. Define a function ε on R by $\varepsilon(\alpha) = 1$ if $\alpha \in R^+$ and $\varepsilon(\alpha) = -1$ if $-\alpha \in R^+$. If $\alpha,\beta \in R$ are linearly independent define

$$\begin{cases} c_{\alpha,\beta} = 0 \text{ if } \alpha+\beta \notin R, \\ c_{\alpha,\beta} = \varepsilon(\alpha)\varepsilon(\beta)\varepsilon(\alpha+\beta)(-1)^{f(\alpha,\beta)} \text{ if } \alpha+\beta \in R. \end{cases}$$

12.13. Lemma. (i) *If* $\alpha,\beta \in R$ *are linearly independent we have* $c_{\alpha,\beta} = -c_{\beta,\alpha}$ *and* $c_{-\alpha,\beta}c_{\alpha,-\alpha+\beta} + c_{\beta,\alpha}c_{-\alpha,\alpha+\beta} = \langle \beta,\alpha^\vee \rangle$;
(ii) *If* $\alpha,\beta,\gamma \in R$ *are linearly independent, we have*
$c_{\alpha,\beta}c_{\alpha+\beta,\gamma} + c_{\beta,\gamma}c_{\beta+\gamma,\alpha} + c_{\gamma,\alpha}c_{\gamma+\alpha,\beta} = 0$.

The first relation of (i) follows from $f(\alpha,\beta) + f(\beta,\alpha) \equiv (\alpha,\beta) \equiv 1 \pmod 2$, if $\alpha,\beta,\alpha+\beta \in R$ (see 12.11(i)). To prove the second one first observe that if $\langle \beta,\alpha^\vee \rangle = 0$ we have $c_{-\alpha,\beta} = c_{\beta,\alpha} = 0$ (12.10(i)). Moreover, since all root strings have length $\leqslant 1$ we cannot have both $c_{-\alpha,\beta} \neq 0$, $c_{\beta,\alpha} \neq 0$. Assume, for example, that $c_{-\alpha,\beta} = 0$, $c_{\beta\alpha} \neq 0$. Then

$$c_{\beta,\alpha}c_{-\alpha,\alpha+\beta} = \varepsilon(\alpha)\varepsilon(-\alpha)\varepsilon(\beta)^2\varepsilon(\alpha+\beta)^2(-1)^{f(\beta,\alpha)+f(-\alpha,\alpha+\beta)} = -1 = \langle \beta,\alpha^\vee \rangle.$$

This establishes (i).

To prove (ii), notice that if, say, $c_{\alpha,\beta}c_{\alpha+\beta,\gamma} \neq 0$, we have $(\alpha,\beta) = (\alpha+\beta,\gamma) = -1$. It follows that either $(\alpha,\gamma) = 0$, $(\beta,\gamma) = -1$ or $(\alpha,\gamma) = -1$, $(\beta,\gamma) = 0$. Assume the first case. Then the relation to be proved follows from

$$f(\alpha,\beta) + f(\alpha+\beta,\gamma) + f(\beta,\gamma) + f(\beta+\gamma,\alpha) \equiv$$

$$(\alpha,\beta) + (\alpha,\gamma) + 2f(\beta,\gamma) \equiv 1 \pmod 2.$$

12.14. We now define a Lie algebra $\mathfrak{g}$ over k as follows. $\mathfrak{g} = (X^{\vee} \otimes k) \oplus \sum_{\alpha\in R} ke_{\alpha}$. The Lie algebra product is defined by

$$(*)\quad \begin{cases} [u \otimes 1, e_{\alpha}] = \langle \alpha, u\rangle e_{\alpha} & (u \in X^{\vee}, \alpha \in R), \\ [e_{\alpha}, e_{\beta}] = c_{\alpha,\beta}e_{\alpha+\beta} & (\alpha,\beta \in R \text{ linearly independent}), \\ [e_{\alpha}, e_{-\alpha}] = \alpha^{\vee} \otimes 1. \end{cases}$$

The verification that this defines a Lie algebra structure is straightforward, using the previous lemma and 12.11(iii). It is left to the reader.

If $a \in \mathfrak{g}$ define the linear transformation ad a of $\mathfrak{g}$ by ad $a(b) = [a,b]$. It follows from (*), since root strings have length ≤ 1, that for all $\alpha \in R$,

$$(\text{ad } e_{\alpha})^2(u \otimes 1) = 0, \quad (\text{ad } e_{\alpha})^2 e_{\beta} = 0 \ (\beta \neq -\alpha),$$

$$(\text{ad } e_{\alpha})^2 e_{-\alpha} = -2e_{\alpha}.$$

Define a linear transformation $x_{\alpha}(\xi) \in GL(\mathfrak{g})$ by

$$x_{\alpha}(\xi) = 1 + \xi \text{ad } e_{\alpha} + \tfrac{1}{2}\xi^2(\text{ad } e_{\alpha})^2,$$

$(\alpha \in R, \xi \in k)$. It is clear from the preceding formulas how to interpret the last term if the characteristic is 2.

Let T be the torus with character group Q, and let T act in $\mathfrak{g}$ by

$$\begin{cases} t.(u \otimes 1) = u \otimes 1 & (u \in X^{\vee}) \\ t.e_{\alpha} = \alpha(t)e_{\alpha} & (\alpha \in R). \end{cases}$$

Since the weights of T in $\mathfrak{g}$ span the character group Q of T, we have that T is isomorphic to its image in $GL(\mathfrak{g})$. Identify T with this image. If $\alpha \in R$, let $\alpha^{\vee}: \mathbb{G}_m \to T$ be the homomorphism of algebraic groups such that $x(\alpha^{\vee}(\xi)) = \xi^{\langle x, \alpha^{\vee} \rangle}$ $(\xi \in k^*, x \in Q)$. Then $\alpha^{\vee}$ can be viewed as an element of the dual $Q^{\vee}$ of Q. Furthermore, define $n_{\alpha} \in GL(\mathfrak{g})$ by $n_{\alpha} = x_{\alpha}(1)x_{-\alpha}(-1)x_{\alpha}(1)$ $(\alpha \in R)$.

The linear maps $x_{\alpha}(\xi)$ of $\mathfrak{g}$ are given explicitly by

$$\begin{cases} x_{\alpha}(\xi)(u \otimes 1) = u \otimes 1 - \xi\langle \alpha, u \rangle e_{\alpha} \quad (u \in X^{\vee}), \\ x_{\alpha}(\xi)e_{\beta} = e_{\beta} + \xi c_{\alpha,\beta}e_{\alpha+\beta} \qquad (\beta \neq \pm \alpha) \\ x_{\alpha}(\xi)e_{\alpha} = e_{\alpha},\ x_{\alpha}(\xi)e_{-\alpha} = e_{-\alpha} + \xi\alpha^{\vee} \otimes 1 - \xi^2 e_{\alpha}. \end{cases}$$

Put $n_{\alpha}(\xi) = x_{\alpha}(\xi)x_{-\alpha}(-\xi^{-1})x_{\alpha}(\xi)$ $(\alpha \in R, \xi \in k^*)$. A straightforward computation then shows that $n_{\alpha}(\xi)$ is given by

$$\begin{cases} n_{\alpha}(\xi)(u \otimes 1) = u \otimes 1 - \langle \alpha, u \rangle (u^{\vee} \otimes 1) = (s_{\alpha}u) \otimes 1, \\ n_{\alpha}(\xi)e_{\alpha} = -\xi^{-2}e_{-\alpha},\ n_{\alpha}(\xi)e_{-\alpha} = -\xi^2 e_{\alpha}, \\ n_{\alpha}(\xi)e_{\beta} = e_{\beta} \text{ if } (\beta,\alpha) = 0, \\ n_{\alpha}(\xi)e_{\beta} = \xi c_{\alpha,\beta}e_{\alpha+\beta} \text{ if } \alpha+\beta \in R, \\ n_{\alpha}(\xi)e_{\beta} = -\xi^{-1}c_{-\alpha,\beta}e_{-\alpha+\beta} \text{ if } -\alpha+\beta \in R. \end{cases}$$

We use here that $c_{\alpha,\beta}c_{-\alpha,\alpha+\beta} = 1$ if $\alpha,\beta,\alpha+\beta \in R$.
Fix a system of positive roots R^+, with basis D, and let G be the subgroup of $GL(\mathfrak{g})$ generated by T, the $x_\gamma(\xi)$ with $\xi \in k, \gamma \in R^+$, and the n_α with $\alpha \in D$.

12.15. _Theorem_. G _is a connected semi-simple adjoint linear algebraic group_, T _is a maximal torus of_ G _and the root system of_ (G,T) _is_ R.
By 12.7 this will follow if we show that the conditions of 12.6 are satisfied. This is a straightforward consequence of the formulas given in 12.14. We leave the details of the verification to the reader.

Remark. The verification will show that the structure constant $c_{\alpha,\beta;1,1}$ equals $c_{\alpha,\beta}$. Hence 12.12 gives an explicit description of structure constants, in the simply laced case. This description is, essentially, due to Frenkel and Kac (see [17 , Prop.2.2]).

12.16. Let G and T be as in 12.15. Fix a system of positive roots R^+ of the root system R, with basis D. Let $\mathcal{D}$ be the corresponding Dynkin diagram (11.3.4) and assume that f is an automorphism of $\mathcal{D}$ with the property (c) of 11.4.3. These automorphisms were described in 11.4.5, and one checks that they have the following property: Let D' be the set of elements of D which are not fixed by f. There is a subset D_1' of D' such that D' is the disjoint union $D' = D_1' \cup fD_1' \cup \ldots$ of f-transforms of D_1', and such that these transforms are mutually orthogonal.

Also, if R is irreducible, the order h of f is either 2 or 3, the second alternative occurring only when R is of type D_4.

The automorphism f of the basis D of the vector space V defines a linear map of V, also denoted by f. It stabilizes R and the root lattice Q. We may also assume that our linear form (,) is f-invariant (because f and W generate a finite group of linear transformations of V, compare 9.1.8). We assume (12.12) that $(\alpha,\alpha) = 2$ for all $\alpha \in R$. Then (x,x) is an even integer if $x \in Q$.

12.17. Lemma. (i) $(1-f)Q \cap R = \phi$;

(ii) <u>If</u> $\gamma,\delta \in R$, $\gamma \neq \delta$ <u>and</u> $\gamma-\delta \in (1-f)Q$ <u>then</u> $(\gamma,\delta) = 0$ <u>and</u> δ <u>lies in the f-orbit of</u> γ.

(i) follows from the fact that if $x = \sum_{\alpha \in D} x_\alpha \alpha$ is a nonzero element of $(1-f)Q$, the nonzero coefficients x_α cannot all have the same sign. We first prove (ii) in the case that the order h of f equals 2. If γ and δ are as in (ii) and D_1' is as in 12.16 we can write $\gamma-\delta = y-fy$, with $y = \sum_{\alpha \in D_1'} x_\alpha \alpha$, whence $(\gamma-\delta,\gamma-\delta) = (y-fy,y-fy) = 2(y,y)$. But $(\gamma-\delta,\gamma-\delta) = (\gamma,\gamma) + (\delta,\delta) - 2(\gamma,\delta) = 4 - 2(\gamma,\delta)$, so $(y,y) = 2 - (\gamma,\delta)$. Since $(\gamma,\delta) = 0,\pm 1$ and (y,y) is even we have $(\gamma,\delta) = 0$. We have, in particular, $(\gamma,f\gamma) = 0$. Then $(\delta,f\gamma) = (\delta-\gamma,f\gamma) = (fy-y,f\gamma) = (y-fy,\gamma) = (\gamma-\delta,\gamma) = 2$. It follows that $(\delta-f\gamma,\delta-f\gamma) = 0$, whence $\delta = f\gamma$. This proves (ii) in the case $h = 2$. To finish the proof of (ii), it suffices to deal with the case that R is irreducible of type D_4, and f is as in 11.4.5, of order 3. The explicit description of the root systems of type D_4 (see 9.3.9(5)) shows that the simple roots can be taken to be the following vectors in the Euclidean space $\mathbb{R}^4$ ((ε_i) is the canonical basis): $\alpha_1 = \varepsilon_1-\varepsilon_2$, $\alpha_2 = \varepsilon_2-\varepsilon_3$, $\alpha_3 = \varepsilon_3-\varepsilon_4$, $\alpha_4 = \varepsilon_3+\varepsilon_4$, f permuting cyclically $\alpha_1,\alpha_3,\alpha_4$. The positive roots are $\alpha_1+\alpha_2$, $\alpha_2+\alpha_3$, $\alpha_2+\alpha_4$, $\alpha_1+\alpha_2+\alpha_3$, $\alpha_1+\alpha_2+\alpha_4$, $\alpha_2+\alpha_3+\alpha_4$, $\alpha_1+\alpha_2+\alpha_3+\alpha_4$, $\alpha_1+2\alpha_2+\alpha_3+\alpha_4$. The assertions of (ii) can now easily be checked explicitly. We leave the details to the reader.

12.18. If $\mathcal{D}$ and f are as above, we define a Dynkin diagram $\mathcal{D}_f$ as follows. The vertices of $\mathcal{D}_f$ are the f-orbits in D. If $\mathcal{O}$ is such an orbit, write $\alpha_{\mathcal{O}} = \sum_{\alpha\in\mathcal{O}} \alpha$. Then two distinct orbits $\mathcal{O},\mathcal{O}'$ are endpoints of an edge if and only if $(\alpha_{\mathcal{O}},\alpha_{\mathcal{O}'}) \neq 0$. If this is so, they are joined by one bond if $\mathcal{O}$ and $\mathcal{O}'$ have the same number of elements, and by h (= 2,3) bonds if $\mathcal{O}$ has one element and $\mathcal{O}'$ h elements, with an arrow pointing towards $\mathcal{O}'$.

From the information given in 11.4.5 it is easily seen that $\mathcal{D}_f$ is indeed a Dynkin diagram. For irreducible root systems, $\mathcal{D}_f$ is given in the following list.

type $\mathcal{D}$		type $\mathcal{D}_f$
A_{2l+1}	(h = 2)	C_l
D_l	(h = 2)	B_l
E_6	(h = 2)	F_4
D_4	(h = 3)	G_2.

We say that $\mathcal{D}_f$ has been obtained from $\mathcal{D}$ by <u>folding</u> (according to f).

We have seen in 11.4.4 that there exists an automorphism ϕ of order h of the algebraic group G of 12.15, which fixes T and induces the automorphism f of Q. Put $G_\phi = \{g \in G \mid \phi g = g\}$, this is a closed subgroup of G. Let $T_\phi = G_\phi \cap T$.

12.19. <u>Theorem</u>. <u>G_ϕ^0 is a connected semi-simple adjoint group. T_ϕ is a maximal torus of G_ϕ^0 and the root system $R(G_\phi^0, T_\phi)$ has Dynkin diagram $\mathcal{D}_f$.</u>

Put $H = G_\phi^0$. We have $T_\phi = \{t \in T \mid \phi t = t\}$. It is easily seen, by considering the homomorphism $t \mapsto \phi(t)t^{-1}$, that

$$\{x \in Q \mid x(T_\phi) = 1\} = (1-f)Q.$$

Since $Q/(1-f)Q$ has no torsion, because f permutes the elements of a basis of Q, it follows that T_ϕ is a torus. One also sees that the Lie algebra of T_ϕ is the set of fixed points of $d\phi$ in $\mathfrak{t}$ = Lie T.

From 12.17(i) we see that no root $\alpha \in R$ is trivial on T_ϕ. This shows that T_ϕ is a regular subtorus of T (see 9.1). Consequently the centralizer $Z_H T_\phi$ lies in $T \cap H = T_\phi$, which implies that T_ϕ is a maximal torus of H.

Let $\alpha \in R$. If $f\alpha = \alpha$ then $Z_G(\mathrm{Ker}\ \alpha)^0$ is a reductive subgroup of G, which is ϕ-stable. The same is true for its commutator subgroup C (isomorphic to $\mathbb{SL}_2$ or $\mathbb{PSL}_2$). Our assumptions imply that ϕ induces an automorphism of C which is the identity on a Borel subgroup of C. Hence this induced automorphism is the identity (7.2.11(8)). We conclude that $H \cap Z_G(\mathrm{Ker}\,\alpha)^0$ is non-solvable, hence the restriction $\alpha|T_\phi$ is a root of (H,T_ϕ).

We next establish the same fact if $f\alpha \neq \alpha$. Let x_α and $x_{-\alpha}$ be isomorphisms $\mathbb{G}_a \to G$ with the usual properties. If $f\alpha \neq \alpha$ one infers from 12.17(ii) that $\xi \mapsto x_{\pm\alpha}(\xi)\phi(x_{\pm\alpha}(\xi))$ (resp. $\xi \mapsto x_{\pm\alpha}(\xi)\phi(x_{\pm\alpha}(\xi))\phi^2(x_{\pm\alpha}(\xi))$ if $h = 3$) defines isomorphisms of $\mathbb{G}_a$ onto closed subgroups of H (recall also that $\phi^h = \mathrm{id}$). One concludes that, again, $H \cap Z_G(\mathrm{Ker}\,\alpha)^0$ is non-solvable.

Denote by N the number of f-orbits in R. What we have established implies that $\dim T_\phi + N \leq \dim H$.

From the decomposition into weight spaces $\mathfrak{g} = \mathfrak{t} \oplus \sum_{\alpha\in R} \mathfrak{g}_\alpha$, using the description of Lie T_ϕ given above, one sees that the fixed point set $\mathfrak{g}_\phi$ of $d\phi$ in $\mathfrak{g}$ has dimension equal to $\dim T_\phi + N$. Since $\mathfrak{h} \subset \mathfrak{g}_\phi$ it follows that $\dim H \leq \dim T_\phi + N$. Consequently $\dim H = \dim T_\phi + N$, and $\mathfrak{h} = \mathfrak{g}_\phi$. It also follows that $\mathfrak{h}$ is spanned by the Lie algebra of T_ϕ and 1-dimensional weight spaces, whose weights are the different roots of (H,T_ϕ). This shows that the unipotent radical of H must be trivial. Hence H is reductive.

Let $R' = R(H,T_\phi)$. For any f-orbit $\mathcal{O}$ denote by $\tilde{\mathcal{O}}$ the image in the character group $Q/(1-f)Q$ of T_ϕ of some element $\alpha \in \mathcal{O}$. Then these elements $\tilde{\mathcal{O}}$ form a basis of R'. Moreover, the dual of $Q/(1-f)Q$ is the fixed point set $Q_f^\vee$ of f in the dual $Q^\vee$ of Q. The elements $\tilde{\mathcal{O}}^\vee = \sum_{\alpha\in\mathcal{O}} \alpha^\vee$ are coroots corresponding to the simple roots $\tilde{\mathcal{O}}$, and $\langle \tilde{\mathcal{O}}\, \tilde{\mathcal{O}}_1^\vee \rangle = \langle \alpha, \sum_{\alpha_1\in\mathcal{O}_1} \alpha_1^\vee \rangle$, if $\alpha \in \mathcal{O}$.

From this information one sees that $\mathcal{D}_f$ is the Dynkin diagram of R', defined by our basis. Since the $\tilde{\mathcal{O}}$ span the character group of T_ϕ, the group H is indeed semi-simple adjoint. We have proved the assertions of 12.19.

12.20. We can now prove the existence theorem 12.1. Let X be as in the statement of the theorem. Denote again by $Q \subset X$ the root lattice, spanned by R and put $X_0 = \{x \in X | \langle x, R^\vee \rangle = 0\}$. Then $Q \oplus X_0$ is a sublattice of X of finite index (9.1.10(1b)). By 12.8 it suffices to consider the case that $X = Q \oplus X_0$. This case is easily reduced to the cases $X = Q$ or $X = X_0$. In the second situation $R = \phi$ and G is a torus (see the beginning of 12.3). If $X = Q$ and if R is simply laced the existence theorem is proved in 12.15. In the general case it follows from the discussion of 12.19, using the classification of irreducible root systems, that we can obtain the Dynkin diagram of any irreducible root system which is not simply laced as a $\mathcal{D}_f$. Then 12.19 implies the existence of semi-simple adjoint groups, with arbitrary root system, which is what we wanted.

12.21. Exercise. Let G be a connected reductive linear algebraic group, and T a maximal torus. Denote the corresponding root datum by $(X, R, X^\vee, R^\vee)$.

(a) There exists a connected reductive group $G^\vee$ over k, together with a maximal torus $T^\vee$, such that the root datum $\Psi(G^\vee, T^\vee)$ is $(X^\vee, R^\vee, X, R)$.

$G^\vee$ is the <u>dual</u> of G. It is unique up to isomorphism. We have $(G^\vee)^\vee \simeq G$.

(b) If G is semi-simple and simply connected (resp. adjoint) then $G^\vee$ is semi-simple and adjoint (resp. simply connected).

(c) Determine the duals of $\mathbb{GL}_n$, $\mathbb{SL}_n$, $\mathbb{SO}_n$, $\mathbb{S}p_{2n}$.

Notes.

An existence proof for a reductive group with given root datum is given in [14 , exp.XXV], even for reductive group schemes. A proof for the semi-simple case is contained in [34 , §5].

The proof of [14] uses Weil's theorem about enlarging an algebraic group germ to an algebraic group, and that of [34] needs results about semi-simple Lie algebras and their representations.

We only use here the adjoint representation of a reductive group, in the case that the root system is simply laced . In order to deal with it we need to prove the existence of semi-simple Lie algebras with a given simply laced root system. Thanks to the remark of Frenkel and Kac quoted after 12.15 the latter existence proof is easy.

For the proof of the main theorem 12.1 of this chapter we also need a result (12.8) about existence of covering groups and explicit information about automorphisms of Dynkin diagrams of simply laced root systems.

The existence theorem of [14] gives the existence of reductive group schemes over $\mathbb{Z}$, with a given root datum. It is useful to have this theorem available, even if one is only interested in algebraic groups over fields. The method of proof of theorem 12.1 used here can perhaps also be used in the case of group schemes over $\mathbb{Z}$.

Bibliography.

For a more complete bibliography about linear algebraic groups see [23].

[1] H. Azad, Structure constants of algebraic groups, preprint.

[2] A. Borel, Groupes linéaires algébriques, Ann. of Math. 64 (1956), 20-82.

[3] A. Borel, Linear algebraic groups, Benjamin, New York, 1969.

[4] A. Borel, Automorphic L-functions, Proc. Symp. Pure Math. 33, part 2, 27-61, Amer. Math. Soc., 1979.

[5] A. Borel and T.A. Springer, Rationality properties of linear algebraic groups II, Tôhoku Math. J. 20 (1968), 443-497.

[6] A. Borel et al., Algebraic groups and related finite groups, Lect. Notes in Math. no. 131, Springer-Verlag, 1970.

[7] N. Bourbaki, Algèbre commutative, Hermann, Paris, 1961-1965.

[8] N. Bourbaki, Algèbre de Lie, Hermann, Paris, 1971-1975.

[9] P. Cartier and J. Tate, A simple proof of the main theorem of elimination theory, L'Enseignement mathématique (II) 24 (1978), 311-317.

[10] C. Chevalley, Fondements de la géometrie algébrique, Paris, 1958.

[11] C. Chevalley, Classification des groupes de Lie algébriques, Séminaire Ecole Normale Supérieure, Paris, 1956-1958.

[12] C.W. Curtis, Central extensions of groups of Lie type, J. reine u. angew. Math. 220 (1965), 174-185.

[13] M. Demazure, Désingularisation des variétés de Schubert, Ann. Ec. Norm. Sup. (4) 7 (1974), 53-88.

[14] M. Demazure and A. Grothendieck, Schémas en groupes, Lect. Notes in Math. nos. 151, 152, 153, Springer-Verlag, 1970.

[15] P. Deligne and G. Lusztig, Representations of reductive groups over finite fields, Ann. of Math. 103 (1976), 103-161.

[16] J. Dieudonné, Cours de géometrie algébrique, 2, Presses Univ. de France, Paris, 1974.

[17] I.B. Frenkel and V. Kac, Basic representations of affine Lie algebras and dual resonance models, to appear.

[18] R. Godement, Théorie des faisceaux, Hermann, Paris, 1958.

[19] D. Gorenstein, Finite groups, Harper & Row, New York, 1968.

[20] A. Grothendieck and J. Dieudonné, Eléments de géometrie algébrique, Publ. Math. I.H.E.S. 1960-1967.

[21] R. Hartshorne, Algebraic geometry, Graduate Texts in Mathematics no. 52, Springer-Verlag, 1977.

[22] J.E. Humphreys, Introduction to Lie algebras and representation theory, Graduate Texts in Math. no. 9, Springer-Verlag, 1972.

[23] J.E. Humphreys, Linear algebraic groups, Graduate Texts in Math., no. 21, Springer-Verlag, 1975.

[24] N. Jacobson, Lie algebras, Interscience, 1962.

[25] N. Jacobson, Basic algebra I, W.H. Freeman and Co., San Francisco, 1974.

[26] S. Lang, Algebra, Addison-Wesley, 1965.

[27] M. Lazard, Sur les groupes de Lie formels à un paramètre, Bull. Soc. Math., France 83 (1955), 251-274.

[28] D. Mumford, Introduction to algebraic geometry, Notes Harvard University.

[29] D. Mumford, Abelian varieties, Oxford University Press, 1970.

[30] I.R. Shafarevich, Basic algebraic geometry, Grundlehren d. Math. Wiss., Bd. 213, Springer-Verlag, 1974.

[31] T.A. Springer, Reductive groups, Proc. Symp. Pure Math. 33, part 1, 3-27, Amer. Math. Soc., 1979.

[32] R. Steinberg, Générateurs, relations et revêtements de groupes algébriques, Colloq. sur la théorie des groupes algébriques, Bruxelles, 113-127, Centre Belge de Recherches Mathém., 1962.

[33] R. Steinberg, Endomorphisms of linear algebraic groups, Mem. Amer. Math. Soc. 80, 1968.

[34] R. Steinberg, Lectures on Chevalley groups, Yale Univ., 1968.

[35] J. Tits, Normalisateurs de tores I. Groupes de Coxeter étendus, J. Alg. 4 (1966), 96-116.

[36] A. Weil, Foundations of algebraic geometry, revised ed., Amer. Math. Soc., Providence, 1962.

Subject Index